SMP 11-16

Book G4

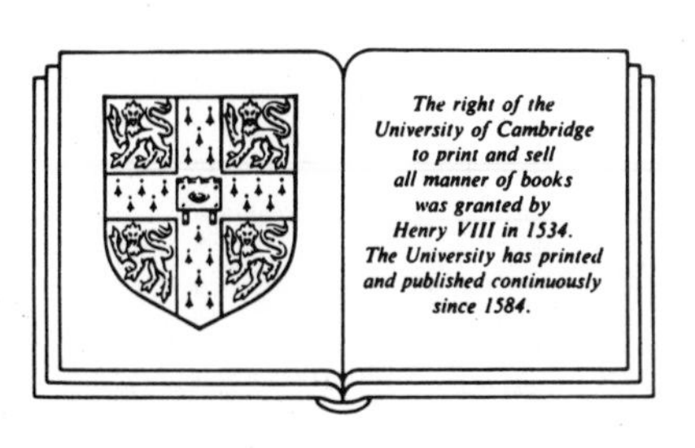

Cambridge University Press

Cambridge
New York Port Chester
Melbourne Sydney

Published by the Press Syndicate of the University of Cambridge
The Pitt Building, Trumpington Street, Cambridge CB2 1RP
40 West 20th Street, New York, NY 10011, USA
10 Stamford Road, Oakleigh, Melbourne 3166, Australia

First published 1986
Fourth printing 1989

Printed in Great Britain at the University Press, Cambridge
Diagrams by Chris Evans and Parkway Group, London and Abingdon
Illustrations by David Parkins, Chris Evans and David Mostyn
Photographs by A. F. Kersting p.44 left and Stanley Freese p.44 centre and right.
Cover illustration by Richard Bonson

British Library cataloguing in publication data
SMP 11–16
Bk. G4
1. Mathematics – 1961–
510 QA39.2

ISBN 0 521 31662 6

Contents

Grid game

GAME FOR 2 Players

You need a dice and 2 counters.

One of you is player A,
the other is player B.
Start in the A and B circles.
Player A starts the first game.
Take turns to throw the dice.
Move your counters using this table.

Dice score	A moves	B moves
1, 2, 3 or 4	1 spot **up**	1 spot **down**
5 or 6	1 spot **right**	1 spot **left**

First to the circles in the centre wins.

The winner gets the points in the circle they land in.

Play again. Take turns to start.
Keep a record of your points.

Game	A's points	B's points
1		
2		

Play the game several times.
Is it better to start at A or B?

Why do you think this is?

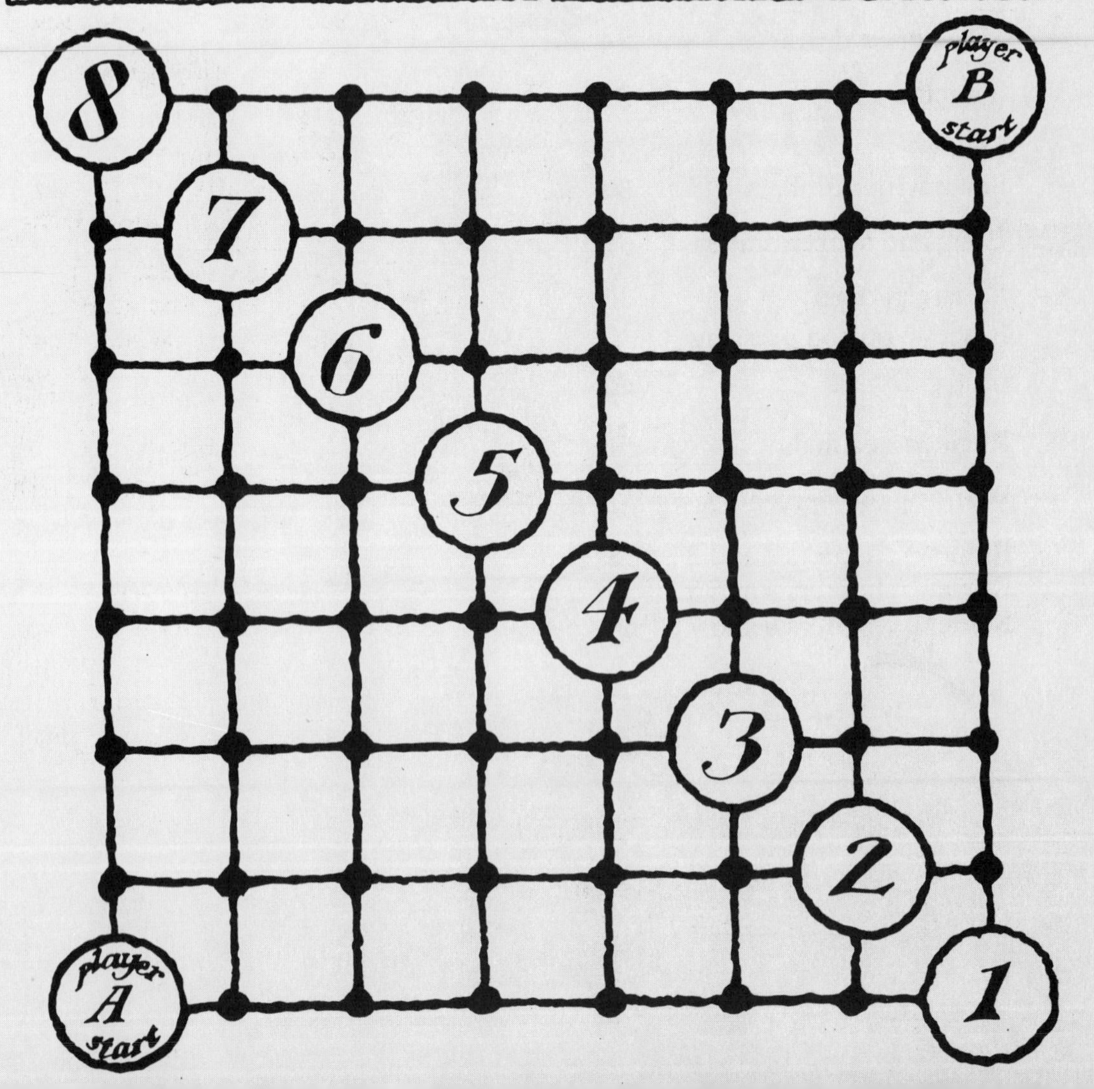

1 John's bike

A Does anyone want a bike?

A1 Make a list of the ways John could try to sell his bike.

Which way do you think will get him the best price?

John decides to put a small advertisement in the local paper. First, he wants to get an idea of how much his bike is worth. He looks through some adverts.

BICYCLE girl's Raleigh 31, excellent condition, £42; Grandstand video game and three cartridges, £40; Sandwich toaster, little used, £10; Mynah bird cage, £6; Tel. Biddle 592415.
BICYCLE girl's, suit 7/10 years; excellent condition; gears, saddlebag; £30 o.n.o. Tel. 437126.
BICYCLE Raleigh Commando, black, £30 – Tel. 512379.
BICYCLE suit boy 9/11 years, little used, 20in. wheels, fitted lock; £35 – Tel. 414379.
BICYCLES Peugeot five-speed Racer, suit 11–14 years, all acccessories, hardly used; £50. Waymaster, suit 5–8 years, £15 – Tel. Napsworth 371.
BICYCLE Raleigh Bomber, very good condition, £40. Electronic chess, as new £25. Tel. 503049.
BLACK vinyl 3-piece suite; good condition, hardly sat on, £45 o.n.o. Tel. 113776.
FOR SALE Gas Fire £60; Copper canopy, £30; two small Fireguards, £6 each. – Tel. 413996.

FOR SALE Gent's 10-speed cycle, alloy chain set; excellent condition, £55; Chopper, fair condition, £20 o.n.o. – Tel. Lower Knowlesworth 3616.
FOR SALE Boys' bikes; Chopper, suit 8–12 yrs; Leopard, suit 7–10 years, £25 each – Call 19, Wayward Close, Clifford Park Estate, evenings.
FOR SALE Full size Space Wars machine; in perfect working order, with coin box mechanism; £250 o.n.o. Tel. 413617 after 4 p.m.
FOR SALE Girl's racing bike; 10 gears; lightweight frame; very good condition; must be seen; offers over £100 – Tel. 615171.
FOR SALE Snooker Table, 2 ft × 4 ft suit 5–8 years, £20; boy's bike, suit 9–12 years, very good condition, £40 – Tel. 316512 after 6 p.m.
FOR SALE Electronic organ: triple keyboard; pedals, memory, drums, auto chords; play to earn money.

GENT'S 24 in. Bike; 5-speed Racer £55, as new. – Tel. 413335.
GENT'S bike in excellent condition; £3[illegible] o.n.o. Also boy's bike – almost brand new £50 o.n.o. – Tel. 525791.
GIRL'S Bike for sale, brand new; £65. – Te[illegible] Broadlands 346.
GIRL'S new racing bike; 10-speed, light weight frame, many accessories, only use[illegible] once; unwanted present; £160 o.n.o. – Te[illegible] 413471.
GIRL'S bike: little used; suit 7 to 8 year old £30; call after 4 p.m. at The Wilding, Flee[illegible] moor Walk, Little Gittingsdale.
GIRL'S and boy's bikes. Both in very goo[illegible] condition, suitable for 10–12 year olds; £4[illegible] each; would consider part exchange fo[illegible] tandem. – Tel. 313919.
GOLD watch for sale; £300 o.n.o. – Te[illegible] 474747.
GRANDFATHER must lose clock, so th[illegible] seven foot clock is now for sale. Only nee[illegible] to be seen. Half-price for quick sale, £25[illegible] no offers.

The adverts John has marked are for bikes like his.

A2 About how much do you think John should ask for his bike?

A3 After some prices in the adverts, you see the letters 'o.n.o.'
(a) Find out what o.n.o. stands for.
(b) What do you think John might get for his bike if he advertised it for '£40 o.n.o.'?

A4 Small adverts cannot tell you everything. Look at this advert for a bicycle.

GIRL'S BIKE suit 8/12 years; Very good condition; £50; Tel. 61651.

Which of these questions can you answer **just** by reading the advert? Write Yes or No for each one.

(a) Is it for a male or a female?
(b) Is it for a child or an adult?
(c) What colour is it?
(d) What make is the bike?
(e) What is the name of the person selling it?
(f) What condition is it in?
(g) How old is it?
(h) What is the address of the person selling it?
(i) What is the phone number of the person selling it?
(j) What is the price?

Option boxes

1 Look through some advertisements.
You will see lots of abbreviations, like o.n.o.

Make a list of abbreviations in the adverts.
Write down what each one stands for.

2 Collect some bicycle adverts from your local papers.
Write out a list of questions like the ones in question A4.
Add some more of your own if you want.

Then make a table to show which questions your adverts answer.

It might start like this.

Advert	Question						
	(a)	(b)	(c)	(d)	(e)	(f)	(g)
RALEIGH Grifter, suit 9/12, red, good condition, £45 – Tel. 451641.	✗	✓	✓	✓	✗	✓	✗

Stick the advert in.

3 This is part of a leaflet that describes John's bike.

BURNER: Rough and tough!

Chrome plated rims, 12 gauge spokes and coloured tyres.

Colour: Flame Red or Blaze Blue

Frame size:
Suitable for legs 22–30 in.

Technical specificiation

WEIGHT	14·5 kg
TYRES	20 × 2·125 20 × 1·75 coloured
SPROCKET	16T Freewhee
CHAIN	88L × ½″ ×

Here is some information about John.

Name John Wright
Address 83, Upper Church St
New Blagdon PG4 3EG
Phone Blagdon 559792
Age 14 years

Make up an advert for John's bike.
Use as much of this information as you want.

B What will it cost?

There is usually a charge for putting an advert in a paper.
Here are the charges for small adverts in a local free paper.
The paper is called *The Advertiser*.

'Trade' is the price they charge to shops and businesses.

'Private' is the rate for ordinary people.

CLASSIFIED CHARGES
10 pence per word (private)
20 pence per word (trade)
Send cheques or P/O to:
Advertiser
47 Millbrook Road
Westhampton
For further information please phone
Tele Sales Dept.
Westhampton 39211

B1 How much would it cost to put each of these adverts in *The Advertiser*?

They are all 'private'.

Abbreviations like o.n.o. or p.m. count as 1 word.

Cycles

RALEIGH Chopper, suit 12 year old, excellent condition, £30. Tel. 416971.

LADY'S cycle, green, panniers and 5 speeds, very good condition. Phone 417942 after 6 p.m.

GIRL'S racing bike, 10 speeds, alloy rims, Reynolds frame, amazingly light, condition as new, £180 o.n.o. Phone 617159.

B2 Look at this bike advert. It is very short!

LADIES vgc, £30, Tel. 466531.

(a) How much would it cost?

(b) Write out each of the adverts in question B1 in a shorter way.

(c) How much do your shorter adverts cost?

Classified Adverts

Only 10p a word.
Minimum charge £2.
Phone 643146

B3 In this paper, each word costs 10p.
But the least you can pay is £2.

(a) How many words do you get for £2?

(b) Write out an advert for John's bike for this paper.

4 John put his advert in the local paper.
Another way would be to put a postcard in a shop window.

Is there a shop near you that does this?
Find out how much a card would cost for a week.

Design a postcard to advertise John's bike.
Make your card as eye-catching as you can.
Use plenty of colour.

Use the details of John and his bike in Option box 3.

5 Think of something you might want to sell one day . . .

1	**Decide how much it is worth.**	How much would it cost new today? Find some adverts from other people selling the same thing. How much are they asking?
2	**Decide how much you are going to ask.**	What sort of condition is it in? Is it a good time of year to sell? Should you ask for a price 'o.n.o.'? How much money do you need?
3	**Work out the cost of your advert.**	Compare different ways – local papers, local radio, shop window and so on. Decide exactly what to put in.
4	**Is it worth it?**	How much money would you make after paying for the advert? Would many people want it? Try doing it!

Oh, £30. But look what I've found in the paper.

Small ads.
GENTS Raleigh racing cycle; 21 inch; 10 speed; immaculate condition; £60 o.n.o. – 1 Ailsa Lane, Bridge Rd., Napsworth.
Smalls wanted: outsize:

2 Thousandths

A Review – hundredths

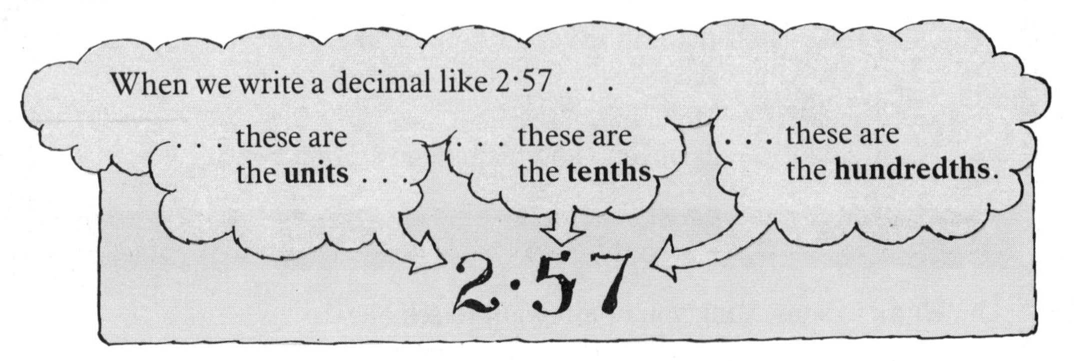

So 1·09 means 1 unit, no tenths and 9 hundredths.
1·90 means 1 unit, 9 tenths and no hundredths.

A1 Write each of these as decimals.

(a) 2 units, 3 tenths and 6 hundredths.

(b) 5 units, 6 tenths and 1 hundredth.

(c) 3 units, no tenths and 5 hundredths.

(d) 3 units, 5 tenths and no hundreths.

A2 We write 4 units and 6 hundreths as 4·06.
The 0 tells us that there are no tenths.
Write each of these as decimals.

(a) 5 units and 8 hundredths. (b) 6 units and 5 hundredths.

(c) 4 units and 1 hundredth. (d) 8 units and 3 hundredths.

When you **say** a decimal, you say each digit separately.
So 4·65 is four point six five **not** four point sixty-five

A3 Write down how you say each of these.

(a) 5·95 (b) 6·70 (c) 9·03

(d) 12·12 (Careful!) (e) 10·10

A4 Count up in hundredths from 1·80 to 2.

'One point eight nought, one point eight one .....'

B Review – m and cm

1 metre is the same as 100 centimetres.
So a centimetre is one hundredth of a metre, 0·01 m.

This is a machine for measuring cloth.
You pull the cloth through it
and it tells you how long the cloth is.

It tells you lengths as
decimals of a metre.
This cloth is 0·46 m long.

B1 How many **cm** long is the cloth?

B2 (a) Is the cloth longer or shorter than $\frac{1}{2}$ m?

(b) By how many centimetres?

(c) What will the machine show when it measures exactly half a metre?

B3 (a) How many cm are there in a quarter of a metre?

(b) What will the machine show when it measures $\frac{1}{4}$ m?

(c) What will it show when it measures $\frac{3}{4}$ m?

B4 Which of these machines is closest to $2\frac{1}{2}$ m?

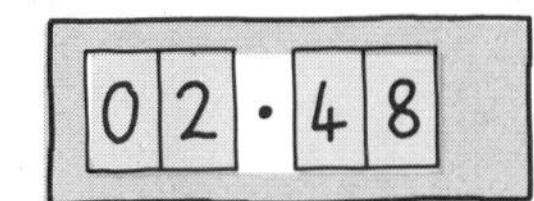

B5 What will the machine show for 3 m 5 cm?

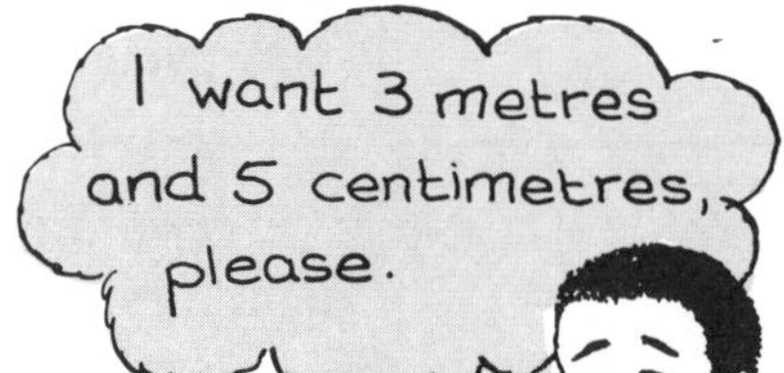

B6 What will the machine show for each of these?

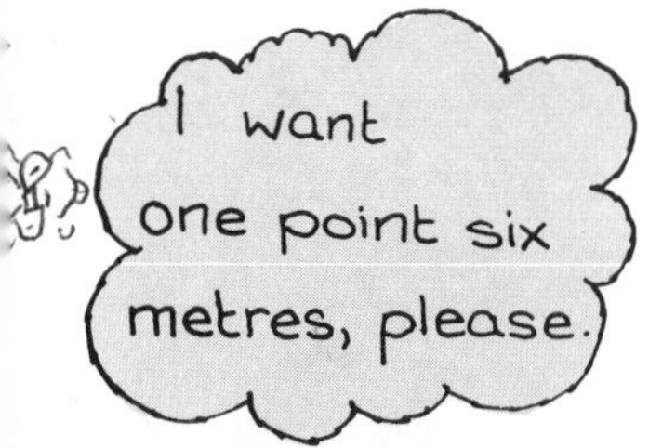

Ten and three quarter metres, please.

One hundred and twenty centimetres, please.

C Kilograms and grams

There are 1000 grams in a kilogram.
So 1 gram is **a thousandth** of a kilogram.

When we write a thousandth in figures, we write it like this.

C1 Write each of these in figures.

(a) Three thousandths (b) Six thousandths

(c) Eight thousandths (d) Nine thousandths

When we write ten thousandths in figures, we write 0·010.
0·010 and 0·01 are the same.

Ten thousandths is the same as one hundredth.

C2 Write each of these in figures.

(a) Twenty thousandths (b) Fifty thousandths

(c) Fifty-five thousandths.

C3 What number is 0·001 more than 0·019?

C4 What number is 0·001 more than each of these

(a) 0·029 (b) 0·059 (c) 0·099

C5 What number is 0·001 **less** than 0·08?
(Remember 0·08 and 0·080 are the same.)

C6 What number is 0·001 less than each of these?

(a) 0·04 (b) 0·01 (c) 0·07

(d) 0·2 (Remember 0·2 and 0·200 are the same.)

C7 Write these numbers in order, smallest first.

0·5, 0·49, 0·51, 0·499, 0·501, 5, 0·05

This weighing machine weighs in kg.
There is 1 gram on the machine.
The weight shows as 0·001 kg.

C8 What will the machine show for 25 grams?

C9 What will the machine show for each of these?

(a)

(b)

(c)

C10 How many grams do each of these weigh?

(a)

(b)

(c)

This flour weighs exactly 1·5 kg.
If you put it on the machine,
the machine shows 1·500 kg

C11 What will the machine show for each of these?

(a)

(b)

(c)

(d) How many grams are there in 0·8 kg

(e) How many grams are there in 0·2 kg

C12 (a) How many grams are there in $\frac{1}{4}$ kg?

(b) What will the machine show for $\frac{1}{4}$ kg?

(c) What will it show for $\frac{3}{4}$ kg?

D Combining

You may need a calculator.

D1 Ajit is weighing this empty beaker.
The beaker weighs 0·096 kg.

(a) How many **grams** does the beaker weigh?

(b) Ajit wants to put exactly 500 g of water in the beaker.

What will the scales read when there is 500 g of water in the beaker?

D2 Here are some objects and their weights.

How much do these weigh?

(a) The flour and the mixing bowl together

(b) 6 eggs

(c) The butter and the jug

(d) The jug with 2 eggs and the flour in it

D3 Which weighs most,
$\frac{1}{2}$ kg of ice cream, 0·6 kg or 380 grams?

D4 A bag of flour weighs 1·5 kg.
You take out 600 g of flour.
How much is left in the bag?

You need a 1·5 kg bag of flour, kitchen scales.

Put the flour on the scales.
Do the scales read more or less than 1·5 kg?
How many grams more or less?

Which do you think is 'right', the flour or the scales?
How could you check? Try it.

E Time

Some races are timed to one thousandth of a second.

E1 (a) This car took 6·471 sec for a race. Is that more or less than $6\frac{1}{2}$ seconds?

(b) If a car's time is less than 6·75 seconds the driver gets a special award. Does this driver get a special award?

(c) The next car took exactly one hundredth of a second less than this one. What was the time of the next car?

E2 How long did Jean take for the race?

E3

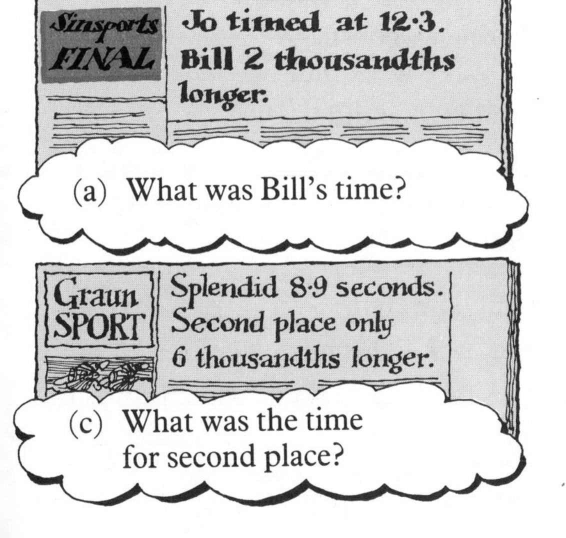

(a) What was Bill's time?

(c) What was the time for second place?

(b) What was Di's time?

(d) What was Chris's time?

Review: graphs

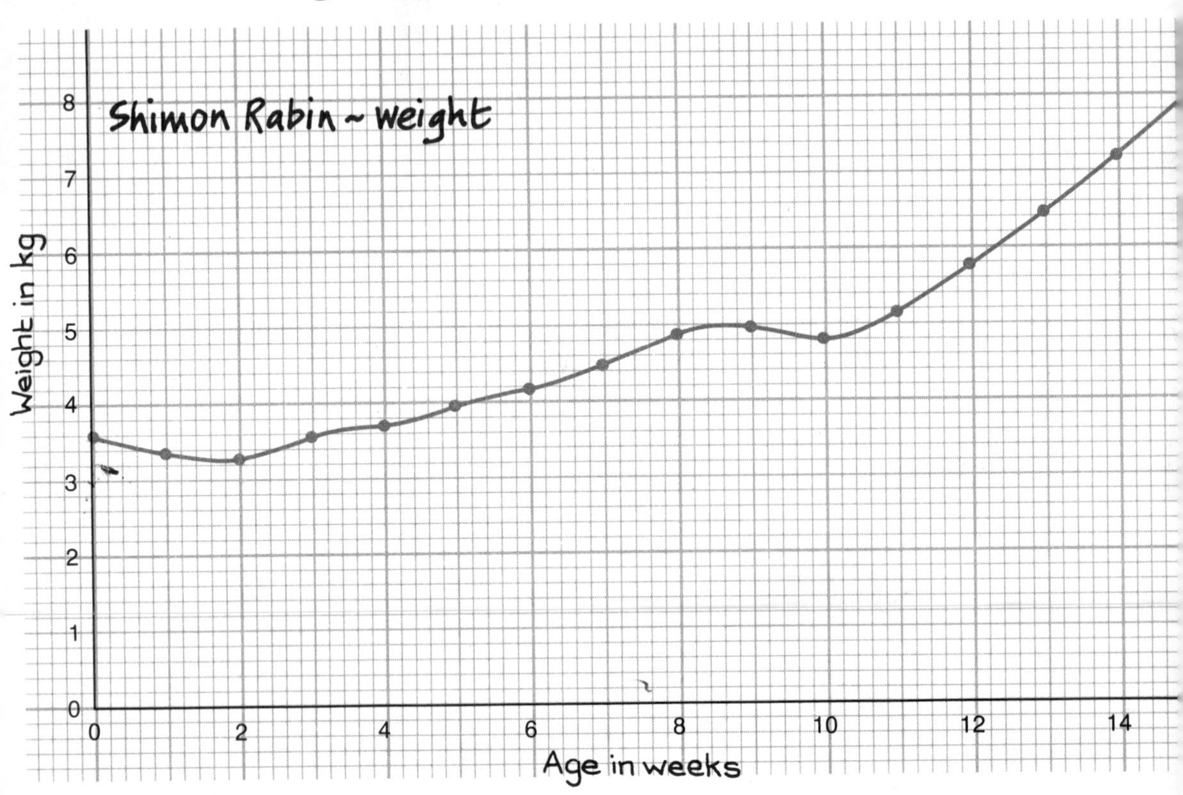

1 This graph shows the weight of a baby boy.
His name is Shimon Rabin.
It shows his weight for the first few weeks of his life.

(a) How many kg did Shimon weigh when he was born? (Write his weight in kilograms, using decimals.)

(b) Shimon lost weight just after he was born. What did he weigh when he was 1 week old?

(c) When Shimon was 2 weeks old, what did he weigh in **kg and grams**?

(d) How many grams did he lose between when he was born and when he was 2 weeks old?

2 It took Shimon a little time to put on weight.

(a) How old was he when he got back to his birth weight?

(b) How old was he when he was **twice** as heavy as when he was born?

3 Shimon was ill for a week while this graph was kept.
How old do you think he was when he was ill?

4 Martina cycles to school every morning.
This graph shows how fast she goes one morning.

For example at 8:38 she was doing exactly 10 m.p.h.

(a) Check that Martina was doing 10 m.p.h. at 8:38.

(b) What time was it when Martina started her cycle ride to school?

(c) After she left home, Martina cycled at a steady speed for a few minutes.
About what was her steady speed?

(d) Martina had to stop at a set of traffic lights.
Roughly when did she stop?

(e) How long did she stop at the traffic lights?

(f) Martina has to cycle up a steep hill on the way to school.
What time do you think she was cycling up the hill?

(g) About how fast did she cycle up the hill.

(h) Martina also cycles down a long hill on the way to school.
What was the fastest she went down hill?

(i) The speed limit is 30 m.p.h.
Martina went faster than this down hill.
For about how long was she breaking the speed limit?

(j) Martina has to be at school by a quarter to nine.
Was she early or late?

3 How many times?

A Competitions

Upper Knowlesworth Silver Band held a sponsored 'blow-in'.
These are the amounts each player raised.

Blow-in : amounts raised

Alf	Beth	Colin	Di	Eric	Fay	Gary	Helen	Ian
£3	£6	£4	£1	£9	£10	£12	£8	£2

A1 Beth got twice as much as Alf.
Who got twice as much as Joy.

A2 There are some other players where one got twice as much as t
Who are they?

A3 Who got 3 times as much as Colin?

A4 Someone got one third of the amount Beth got.
Who was it?

A5 Someone got one fifth of the amount that Joy raised.
Who was it?

A6 Eric raised 3 times what he got in last year's 'blow-in'.
How much did he get last year?

A7 Gary and Helen together raised 5 times as much as
someone else this year.
Who was the 'someone else'?

A8 Eric and his girl friend together got 5 times as much as Alf.
Who is Eric's girl friend?

A9 Altogether the band raised 6 times as much this year
as they raised last year.
How much did they raise last year?

Lower Knowlesworth held a 'welly chucking' competition.
The results were put on this chart.

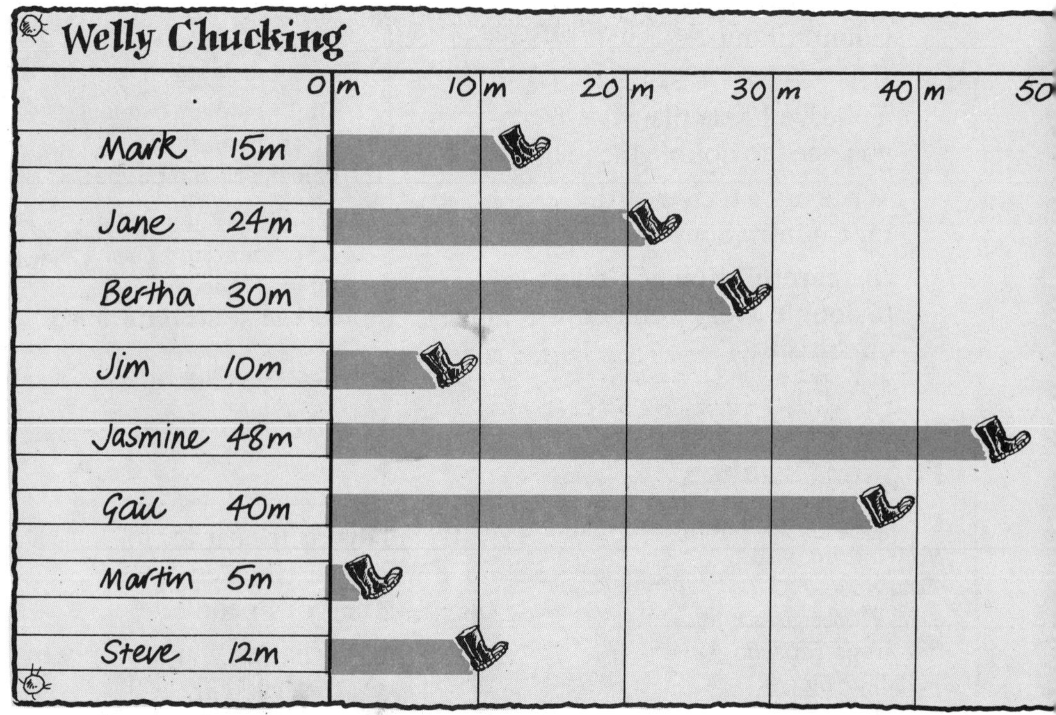

Bertha threw a distance of 30 m.
Martin threw a distance of 5 m.

So you can write

Bertha's distance is 6 times Martin's distance.

A10 Copy and complete these sentences.

(a) *Jane's distance is . . . times Steve's distance.*
(b) *Jasmine's distance is . . . times Steve's distance.*
(c) *Mark's distance is . . . times Martin's distance.*

A11 What is the name of this welly chucker?

My distance is 4 times Jim's.

A12 What are the names of these people?

B Recipes

B1 This is a recipe card for Country Soup.
It serves 6 people.

To serve 12 people you need to double the recipe.

Write out a recipe card for Country Soup for 12 people.

(**Be careful!** You don't have to double **every** number on the card.)

Country Soup — serves 6

225 g	fresh minced beef
3	carrots
2	leeks
2	potatoes
30 ml	medium oatmeal
2 litres	beef stock

bay leaf, salt and pepper

Fry mince 5 minutes.
Add other ingredients.
Simmer 40 minutes.
Serve hot with brown bread.

15ml is about 1 tablespoon.

Pepper Chicken — serves 4

45 ml tomato ketchup
30 ml tomato paste
45 ml vinegar
30 ml Worcester sauce
few drops Tabasco sauce
4 chicken pieces

Place everything except chicken in a roasting pan. Mix well.
Add chicken, leave 2 hours.
Heat oven to Gas Mk 6, 400 °F.
Bake 45 min, or until done.

B2 This recipe card is for Pepper Chicken.
It serves 4 people.

(a) What do you multiply by to serve 12?

(b) Write out a recipe card for Pepper Chicken for 12.

B3 These are the ingredients for rice pudding for 50.

(a) 200 people have rice pudding at a works canteen.
Write down a list of the ingredients needed.

(b) Suppose you make rice pudding for 10 people.
How much custard powder would you need?

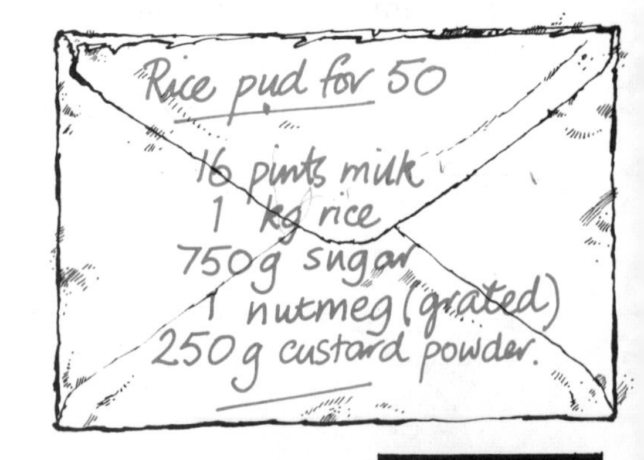
Rice pud for 50

16 pints milk
1 kg rice
750 g sugar
1 nutmeg (grated)
250 g custard powder.

B4 1 pint of milk is enough for about 16 cups of coffee.
How much milk would you need for 48 cups of coffee?

B5 60 grams of tea makes about 30 cups.
How much tea do you need for 10 cups?

B6 You can get about 16 portions from a large packet of cornflakes.

(a) How many breakfasts can 4 people get from one packet?

(b) How many breakfasts can 2 people get from one packet?

(c) 32 people have cornflakes for breakfast for 4 days.
About how many packets will they need?

B7 1 litre of custard serves 9 people.

(a) How much custard would you need for 18 people?

(b) George is making 4 litres of custard for a camp of 40 people.
Do you think he is making enough?

B8 1 litre of soup serves about 4 people.

(a) How many does 5 litres serve?

(b) How much soup would you make to be sure of serving 100 people?

B9 2 kg of mince serves about 18 people.

(a) How much mince do you need for 9 people?

(b) Is 3 kg enough for 25 people?

B10 Here are two lists of ingredients for apple pie.

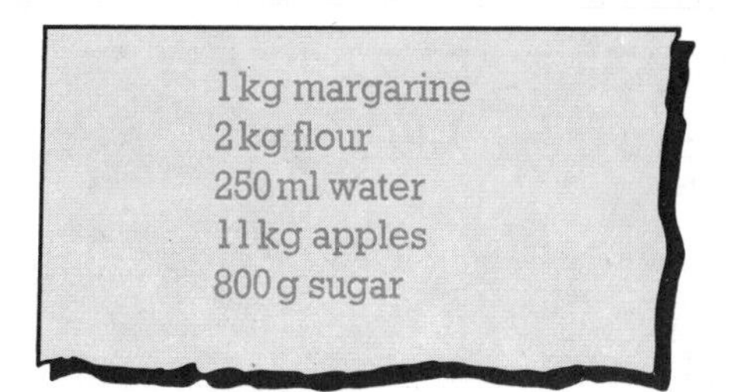

1 kg margarine
2 kg flour
250 ml water
11 kg apples
800 g sugar

This recipe serves 48.

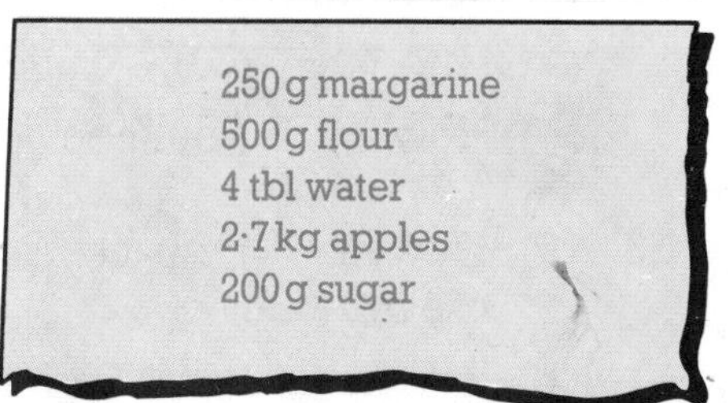

250 g margarine
500 g flour
4 tbl water
2·7 kg apples
200 g sugar

How many does this recipe serve?

C Bigger is cheaper?

Sometimes you can buy the same thing in different sizes.
This margarine comes in 250g tubs, 500g tubs and 2kg tubs.

C1 Suppose you want 2kg of margarine.

(a) How many 500g tubs do you need to make 2kg?

(b) How many 250g tubs make 2kg?

C2 How much does it cost to buy 2kg of margarine in 500g tubs?

C3 How much does 2kg cost in 250g tubs?

C4 What is the cheapest way to buy 2kg of this margarine?

Soap powder is sold in 'Euro-sizes'.
These are Euro-sizes E5, E10 and E20.

In the McGee family
there are 2 adults and 3 children.

The McGees use six E10
packets of soap powder each year.

C5 What weight of soap powder do the McGees use in a year?

C6 The E10 packet costs £2·59.
How much does soap powder cost the McGees each year?

C7 The E20 packet contains twice as much powder as an E10 packet. It costs £3·89.

(a) How many E20's would the McGees use in a year?

(b) How much would powder cost them if they used E20's?

C8 (a) How many E5 packets would the McGees use in 1 year.

(b) The E5 packets are on special offer.
At the moment, an E5 packet costs 95p.
How much would the McGees soap powder cost in a year if they used E5 packets?

C9 Copy and complete the label on the 50 g jar.

C10 About how many cups of coffee would you make from a 200 g jar?

C11 These were the prices in 1982 for one brand of instant coffee. Suppose you want 500 g of coffee.

(a) How much would 10 50 g jars cost?

(b) How much would 500 g cost if you buy it in 100 g jars?

C12 There are lots of different ways of buying 500 g.

(a) List some ways you can buy 500 g.

(b) What is the **cheapest** way to buy 500 g?

C13 About how many cups will 500 g of coffee make?

discussion points

Why are things often cheaper when you buy them 'in bulk'?

Sometimes it is not sensible to buy in bulk.
When is it silly to buy huge packets?

Supermarkets are often cheaper than small shops.
Why do you think this is?

Why do people buy things from small shops?

Is 'cheapest' always 'best'?

4 Timetables

Trains to London

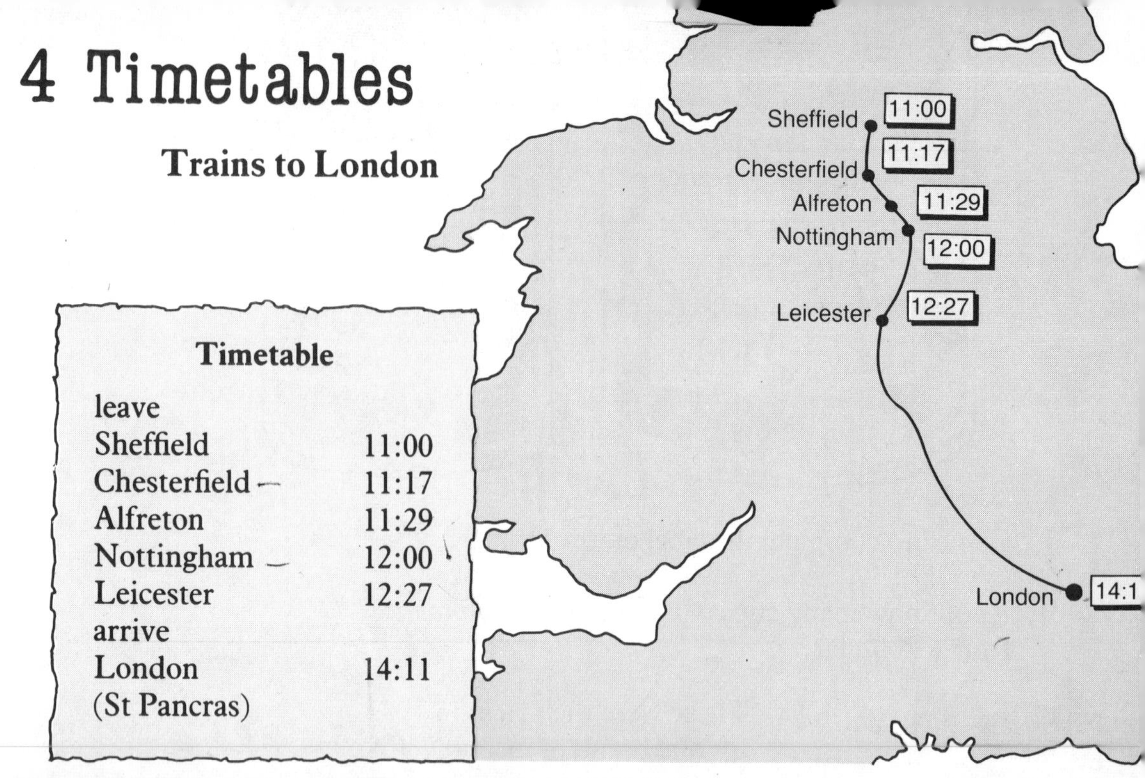

Timetable

leave	
Sheffield	11:00
Chesterfield	11:17
Alfreton	11:29
Nottingham	12:00
Leicester	12:27
arrive	
London (St Pancras)	14:11

A1 Imagine you are on the 11 o'clock train from Sheffield to London. Which stations do you think are shown in each of these pictures? **(Remember trains don't always keep exact time.)**

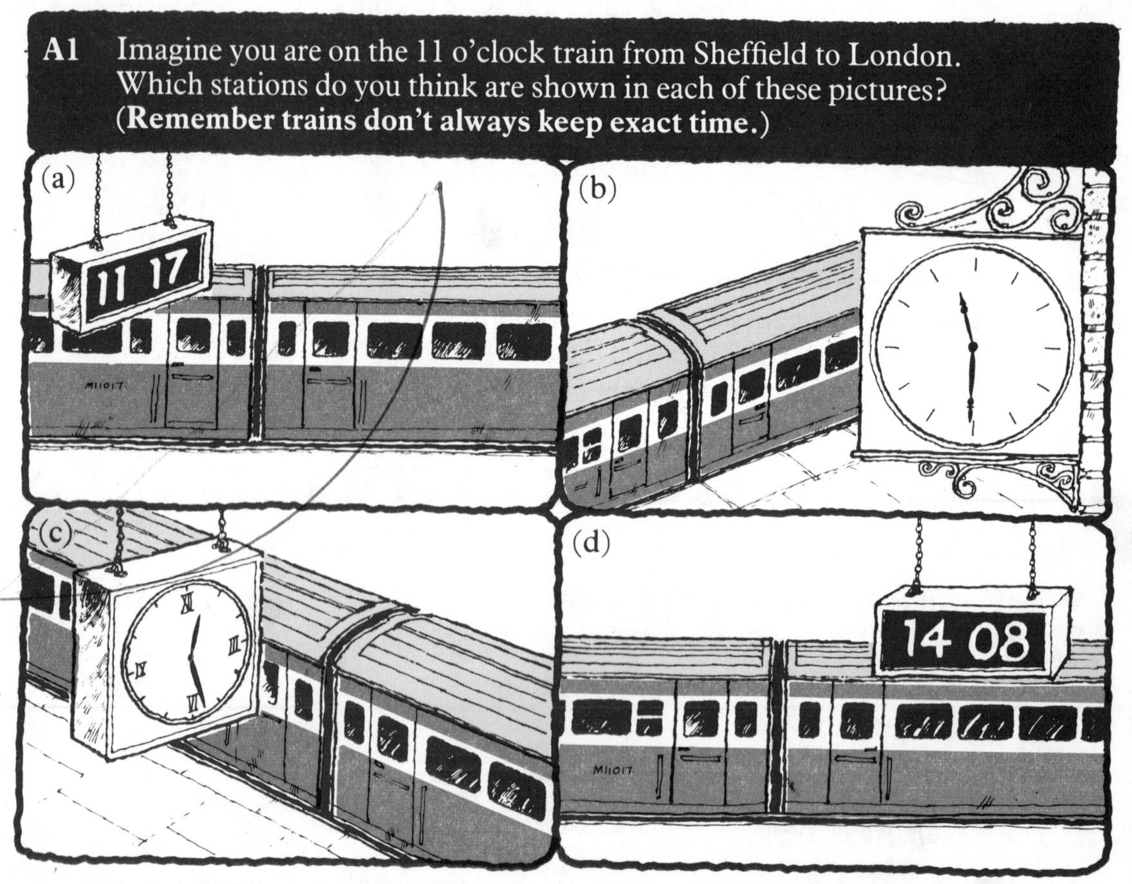

11:00 12:00 13:00 14:00

A2 If the 11 o'clock train from Sheffield keeps to the timetable, how long will it take from

(a) Sheffield to Chesterfield
(b) Nottingham to London
(c) Chesterfield to Nottingham
(d) Leicester to London.

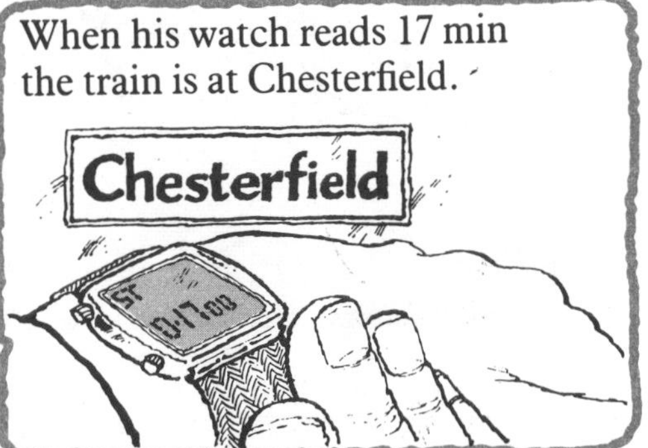

A3 Here are some pictures of Alan's stopwatch on the journey. Which town is the train nearest to in each one?

(a)

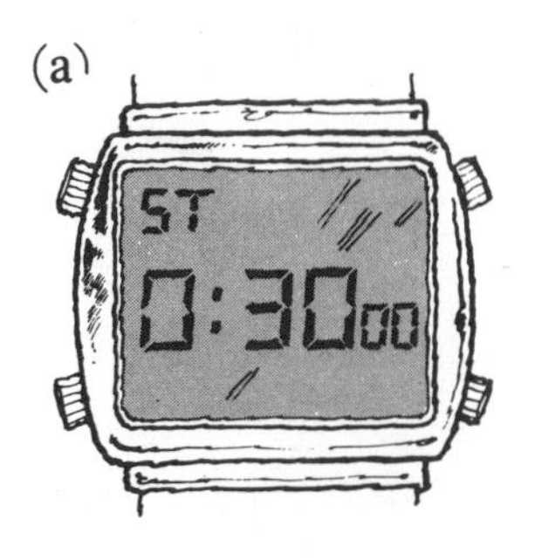

(b)

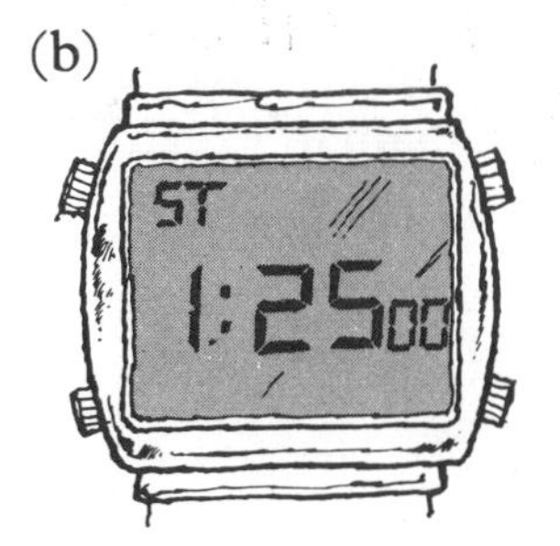

(c)

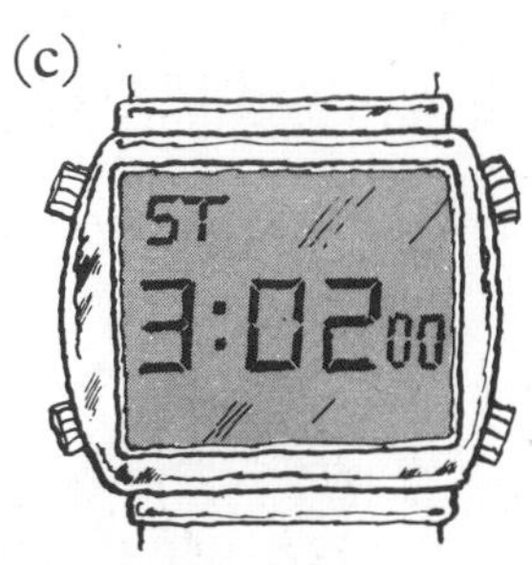

A4 Ann joins the train at Alfreton.
She gets off one hour later.
Where do you think she gets off?

A5 The train gets to London at 14:08.

(a) How many minutes early is it?
(b) Alan is meeting a friend at the station at a quarter to three. How long will Alan have to wait?

This timetable also shows the 11 o'clock train from Sheffield.
It shows two other trains as well.

You read the times **across** this timetable.
The 1 o'clock train departs from Sheffield at 13:00.
The 3 o'clock train departs at 15:00.

depart Sheffield	Chesterfield	Alfreton	Nottingham	Leicester	arrive London (St Pancras)
1100	1117	1129	1200	1227	1411
1300	1317	1330	1400	1427	1610
1500	1517	1528	1555	1620	1750

A6 (a) Each train makes four stops between Sheffield and London. Write down the names of the stops.

(b) Which train is the slowest to Alfreton?

(c) How long does the 1 o'clock train take from Sheffield to London?

(d) How long does the 3 o'clock take from Nottingham to London?

(e) Which train is the fastest from Sheffield to London?

A7 Jane is on the 11 o'clock from Sheffield.

How long is it until her train is due in London?

A8 (a) Ray is on the 1 o'clock. How long until he is due in London?

(b) Leslie is on the 3 o'clock. How long until her train is due at St Pancras?

A9 This train is just arriving at Alfreton.

What time do you think it left Sheffield?

A10 This train is arriving at Leicester.

What time do you think it left Sheffield?

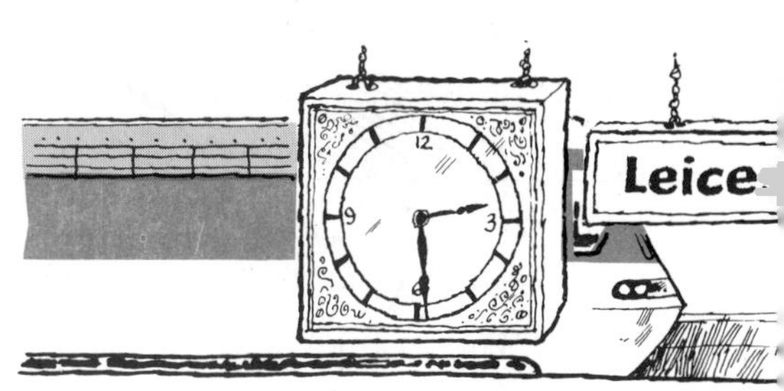

Sheffield	Chesterfield			St Pancras
06.07	06.24	Ⓣ		08.59
07.20	07.36	ǁ	☕	09.50
07.55	08.12	ǁ	☕	10.43
08.55	09.12		☕	12.09
10.12	10.29		☕	13.08
11.00	11.17	ǁ	☕	14.11
12.10	12.27	ǁ	☕	15.13
13.00	13.17	ǁ	☕	16.10
14.10	14.27	ǁ	☕	17.11
15.00	15.17	ǁ	☕	17.50
16.02	→		☕	18.51
17.10	17.27		☕	20.03
17.59	18.16		☕	21.02
19.03	19.20			22.07
20.02	20.19		☕	23.30
23.59	00.20			05.15

For whole or part of journey

ǁ Restaurant service according to the time of day

☕ Drinks and cold snacks

Ⓣ Hot dishes to order; also drinks and snacks

This timetable shows all the trains from Sheffield to London. They all go to St Pancras, London.

It only shows three stations. The others have been left out.

Check that you can find the 11 o'clock, 1 o'clock and 3 o'clock trains.

A11 What does ǁ mean on the timetable?

A12 What do you think → means on the 16:02?

A13 (a) Could you get a cup of coffee on the 16:02?

(b) What is the last train from Sheffield you can get coffee o

A14 (a) How many trains leave Sheffield for London before mid-day?

(b) How long does the 07:20 take from Sheffield to London?

(c) How many trains have 'restaurant service'?

A15 Aneeta lives in Chesterfield.
She wants to get to London at about 8 p.m.
What time train should she catch?

A16 How long does the last train take from Sheffield to London?

B *LinkLine*

The map shows the ***LinkLine*** between Sheffield and Barnsley.

All the lines on the map are railway lines.

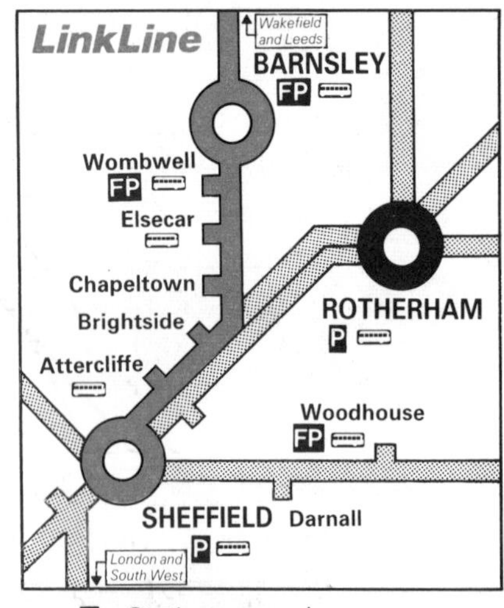

B1 Which places have free car parks at the station?

B2 Which stations have car parks where you have to pay?

B3 Is there a bus service next to Brightside station?

This table shows the single fares in 1982 between places on the ***LinkLine***.

Check that the fare from Brightside to Elsecar was 28p.

	Barnsley	Wombwell	Elsecar	Chapeltown	Brightside	Attercliffe
Wombwell	17p					
Elsecar	20p	16p				
Chapeltown	26p	24p	17p			
Brightside	36p	35p	28p	22p		
Attercliffe	36p	35p	29p	25p	14p	
Sheffield	36p	35p	29p	25p	15p	11p

B4 What was the fare from Sheffield to Elsecar?

B5 How much was it from Barnsley to Chapeltown?

B6 The return fare is twice the single fare.
So a return from Brightside to Elsecar and back was 56p.

(a) How much was a return from Sheffield to Barnsley?

(b) Alan lives in Sheffield and works in Barnsley.
How much did it cost for 5 return tickets each week?

(c) Kate lives in Chapeltown and works in Sheffield.
How much did it cost her to travel by train each week?

B7 Children under 16 travel half fare.
These people are going to Sheffield and back.

What would it have cost altogether in 1982?

Weekly season tickets

Sheffield – Barnsley	£2·40
Sheffield – Chapeltown	£2·10
Sheffield – Elsecar	£2·40
Sheffield – Wombwell	£2·40
Barnsley – Chapeltown	£2·20
Barnsley – Elsecar	£1·60
Barnsley – Wombwell	£1·40

A weekly season ticket between two stations lets you do as many journeys in 1 week as you want.

It is usually cheaper to buy a season ticket if you travel every weekday.

These were the prices in 1982.

B8 Look at your answers to question B6.

(a) Was a weekly season cheaper for Alan?

(b) How much would he save each week?

(c) How much would Kate save each week with a season?

B9 (a) Jo went from Sheffield to Wombwell and back 4 days a week. Which was cheaper, return tickets or a weekly season?

(b) Ruth went from Sheffield to Elsecar and back 4 days a week. Was a season ticket cheaper than 4 return fares?

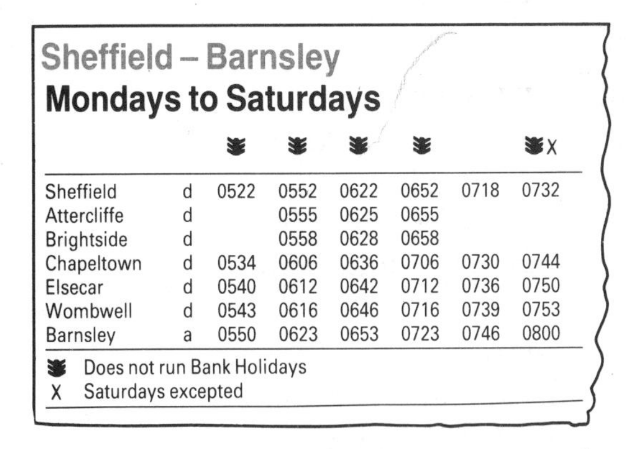

Sheffield – Barnsley
Mondays to Saturdays

		❋	❋	❋	❋		❋X
Sheffield	d	0522	0552	0622	0652	0718	0732
Attercliffe	d		0555	0625	0655		
Brightside	d		0558	0628	0658		
Chapeltown	d	0534	0606	0636	0706	0730	0744
Elsecar	d	0540	0612	0642	0712	0736	0750
Wombwell	d	0543	0616	0646	0716	0739	0753
Barnsley	a	0550	0623	0653	0723	0746	0800

❋ Does not run Bank Holidays
X Saturdays excepted

This shows part of the timetable of trains from Sheffield to Barnsley. It shows the trains early in the morning.

B10 What do you think this 'd' stands for?

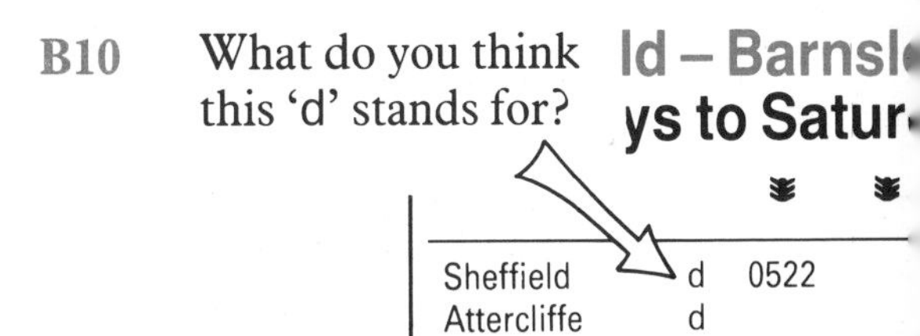

B11 I get to Sheffield at a quarter to six. What time is the first train to Barnsley I can catch?

B12 I want to be in Wombwell before 7 o'clock. What trains can I catch from Attercliffe?

B13 Which is the first train from Sheffield that stops at Brightside?

B14 You have to be in Barnsley just after 8 a.m. Which is the best train from Sheffield on a Saturday?

C Buses

The City Clipper is a bus service round the centre of Sheffield.

It is a free service – you do not have to pay.

The timetable gives these details of buses from Pond Hill.

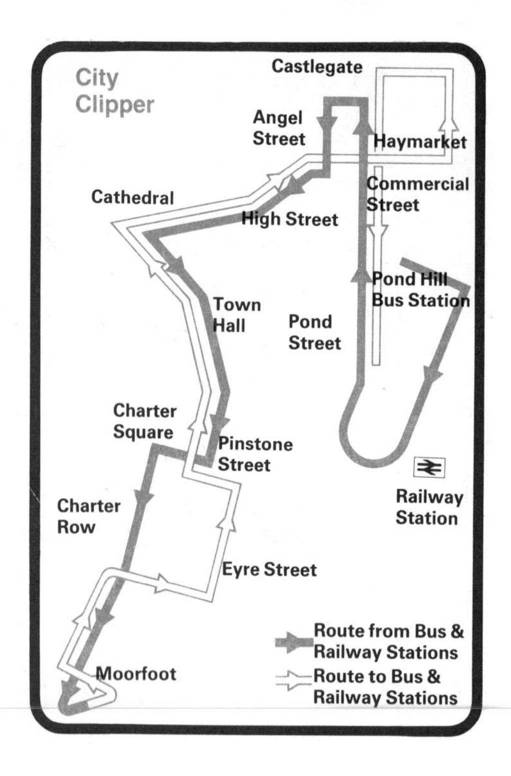

Monday to Friday

POND HILL . . . *dep.* 0930 then approx. every 3 or 4 minutes until 1500, then every 11 minutes until 1730.

Saturday

POND HILL . . . *dep.* 0930 then approx. every 3 or 4 minutes until 1731.

C1 (a) Suppose the buses run every 3 minutes. How many will run each hour?

(b) Suppose the buses run every 4 minutes. How many will run in an hour?

(c) The timetable says 'every 3 or 4 minutes'. **About** how many an hour do you think this is?

(d) About how many buses will run from Pond Hill between 09:30 and 15:00?

C2 After 15:00 the buses run every 11 minutes on weekdays. **About** how many will run from 15:00 to 17:30? (Make a timetable if it helps.)

C3 About how many buses are there altogether from Pond Hill each weekday?

C4 (a) Are there more or less buses from Pond Hill on a Saturday than on a weekday?

(b) About how many buses run from Pond Hill on a Saturday?

C5 There are no 'City Clippers' on Sunday. Why do you think this is?

SHEFFIELD EARLY BIRD

From **Pond Hill Bus Station**

Service operated by
ONE MAN FAREBOX buses

Monday to Friday
POND HILL . . . *dep.* 0745 then every
6 minutes until 0903

FARE 2p

NO CHANGE GIVEN

Holders of British Rail weekly
season tickets may travel free

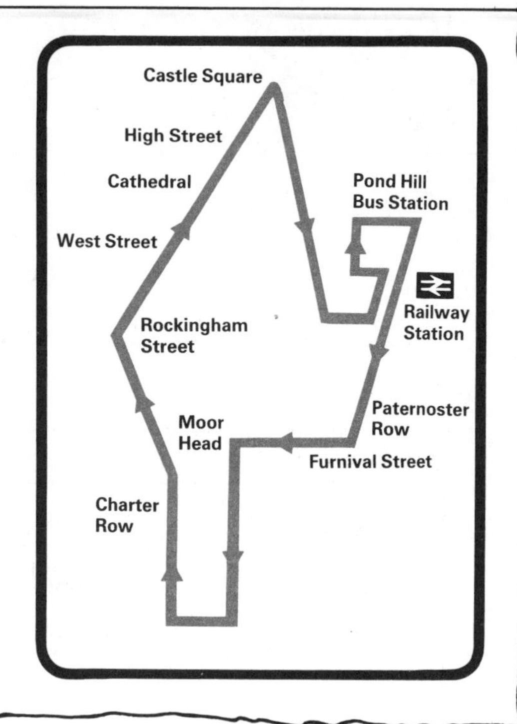

The 'Early Bird' bus runs early in the morning.
Like the 'City Clipper' it also goes round the centre of Sheffield,
but it uses a different route.

C6 About how many 'Early Bird' buses are there each morning?

C7 Each 'Early Bird' bus carries about 40 passengers.

(a) How much money does the driver collect each trip?

(b) About how much money is collected each morning?

(c) About how much do the 'Early Bird' buses collect in fares each week?

C8 There are no 'Early Birds' on Saturdays or Sundays.
Why do you think this is?

discussion points

The 'Early Bird' buses travel 270 miles altogether each week.
In 1982, it cost Sheffield £1·72 for each mile a bus went.
Did Sheffield make a profit on 'Early Birds'?
Did Sheffield make a profit on the 'City Clippers'?
Why do you think Sheffield ran these buses?

You need a dice and worksheet G4–1.

You are a special agent working in Glasgow.

These instructions arrive.

You must go to Morecambe. Then you have to go to Dover and catch a ferry or hovercraft to France.

But trains are always delayed . . .

URGENT

Travel by train to Morecambe.
Collect message in buffet.
Then to Dover and France.
Change trains often.
Rudolph.

Remember

1 You start on the 7:10 from Glasgow. This **should** arrive in Lancaster at 9:29.
But . . .
Trains in this game are always late!

2 To find how late a train will be between two towns

roll the dice.

Multiply the score on the dice by the number in the circle between the two towns.

This tells you how many minutes the train will be late arriving.

3 Use the timetables to decide which trains to catch.

4 You **must** go to Morecambe.

5 It takes you exactly 1 hour to get from London Euston to Charing Cross.

6 It takes you exactly 20 min from Dover station to either the Ferry Port or the Hover Port.

7 Fill in the worksheet as you go.

8 The person who leaves the country earliest is the winner!

Station	Timetable Arrival	Real Arrival	Depart
Glasgow			7 10
Lancaster	9 29	10 19	10 45
Morecambe	10 55		
Lancaster			
Crewe			
Birmingham			

Here is the start of Karen's journey.

She catches the 7:10 from Glasgow.

This should arrive at Lancaster at 9:29.

She throws the dice and gets 5.
So her train is 50 mins late.
She gets to Lancaster at 10:19.

From the timetable, the next train to Morecambe departs at 10:45.

This should arrive at 10:55, and so on.

dep. Glasgow Central	0710	0723	0823	1010	1115	1210	1323	1523	1710	1823
arr. Lancaster	0929	0949	1057	1228	1348	1427	1553	1757	1929	2102

dep. Lancaster	0839	0935	1045	1120	1133	1201	1353	1432	1521
arr. Morecambe	0849	0945	1055	1130	1146	1211	1403	1442	1531
dep. Morecambe	0855	1010	1105	1140	1226	1310	1408	1505	1545
arr. Lancaster	0905	1020	1115	1150	1236	1321	1418	1516	1557

dep. Lancaster	arr. Crewe
0949	1059
1029	1142
1228	1406
1348	1456
1427	1537
1553	1700
1750	1801
1929	2035

dep. Crewe	arr. Birmingham
1104	1200
1341	1405
1406	1500
1501	1607
1550	1700
1713	1820
1753	1900
1822	1923
1915	2025

Birmingham (New Street) to London Euston

dep. B'ham	arr. Euston		
1148	1333	☕	🍴
1218	1405	☕	🍴
1248	1433	☕	🍴
1318	1505	☕	🍴
1348	1533	☕	🍴
1418	1605	☕	🍴
1448	1633	☕	🍴
1518	1702	☕	
1548	1736	☕	🍴
1618	1800	☕	🍴
1648	1830	☕	🍴
1718	1858		
1748	1931	☕	
1818	2003		
1848	2028		
1929	2111	☕	🍴
2003	2143	☕	
2148	2329		

Charing Cross – Dover

dep. Charing Cross	arr. Dover
1800	1941
1900	2023
2000	2128
2100	2229
2130	2314
2230	0015
2303	0054
0628	0826

DOVER FERRY PORT
Sailings to France

1905
2030
2110
2200
2250
2340
0115
0605
0800
0910

DOVER HOVER PORT
Hovercraft to France

depart 0800
1000
1200
1400
1600
1800
2000

Say it with numbers

Try your own starting number.

What seems to happen?

Does it always happen?

Why do you think it happens?

5 Long and short numbers

A Without a calculator

A1 Jane and Marita share £1·20 between them.
They each take the same amount.
How much do they get each?

A2 Glen, Pat and Kevin make money by doing odd jobs.
One Saturday they earn £7·29.
They share it out equally between them.
How much do they each get?

A3 5 friends share some stamps.
There are 305 stamps.
How many stamps does
each friend get?

A4 Sam leaves £4200 in his will.
It is shared equally by 4 charities.
How much does each charity get?

A5 Raffle tickets cost 40p each.
3 friends buy one between them.
How much should they each pay?

discussion points

1 How did you do question A2?
Did everyone in your class do it in the same way?

2 What is the easiest way to work out each of these?

£6 shared between 3 people?	£6·30 shared between 6 people?
£7·50 shared between 3 people?	£7·50 shared between 6 people?

3 What would you do if you were one of the people in question A5?
What is the fairest thing to do?
What do people usually do when money can't
be shared out exactly equally?

B Using a calculator

You need a calculator.

B1 17 people organise a party.
The party costs £213·52 altogether.
How much should each of them pay?

B2 26 people win £10 250·24 on the pools.
They share the money out equally.
How much does each person get?

B3 Zeke has 2 daughters and 4 sons.
He wants to give some money to them.
He has £39 to give away.

(a) If he shares it equally between them all, how much do they each get?

(b) If he gives it just to his daughters, how much do the girls each get?

(c) If he gives it just to his sons how much does each son get?

When you divide £50 by 7 on the calculator it shows

7·1428571

This means £50 won't divide exactly by 7.

B4 Suppose the 7 friends take £60 next week.
Use the calculator to work out
how much they will each get.

B5 Six friends buy a raffle ticket.
They win £50!

They share the £50 equally between them.
How much can they have each?

discussion point

Aunt Aggie sends 50p
to be shared by her nephews.
There are 7 of them!
What should they do?

B6 Kim, Lucy and Keith win £50 in a raffle.
They share the £50 equally between the 3 of them.
How much do they get each?

B7 16 friends win the pools!
The cheque arrives. It is for £15!
How much do they get each?

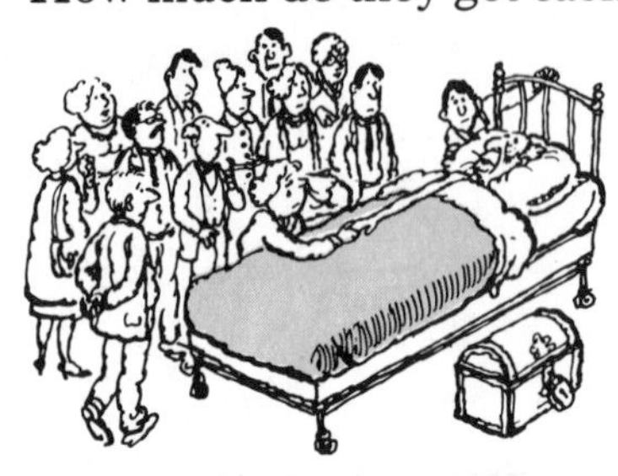

B8 When old George died, he left his money to be shared equally by his 13 children.
When they opened his savings they found £365.
How much should they get each?

B9 In each of these, work out how much each person gets.
(a) £60 is shared between 13 people.
(b) £120 is shared by 7 people.
(c) £1000 is shared by 12 people.
(d) £3500 is shared by 3 people.

B10 3 friends share some matchbox labels.
There are 740 labels.
How many can they have each?

B11 (a) 9 artists share 130 tubes of paint.
How many tubes each can they have?
(b) The same artists share 65 brushes.
How many brushes can they each have?

discussion point

7 friends share £20 between them.
How much do they get each?

7 friends have a meal together.
The meal costs £20.
How much do they each have to pay?

Why should the two answers be different?

C Lengths

This line is 6 cm 5 mm long.

Measure the line and check.

You can write 6 cm 5 mm as a **decimal of a cm**.

6·5 cm

These are the **cm**.

These are the **tenths** of a cm.

C1 Measure each of these lines.
Write the lengths as decimals of a cm.

(a)

(b)

(c)

Nicoletta wants to cut this rod into 3 equal pieces.

First she measures it. It is 8·5 cm long.
Measure the rod yourself and check.
Nicoletta uses a calculator to work out 8·5 ÷ 3.

2.8333333

These are the **cm**.

These are the **tenths** of a cm or mm.

These are **smaller** parts of a cm.

So each part she cuts needs to be about 2·8 cm or 2 cm 8 mm long.

C2 (a) Measure this rod.

(b) You want to cut the rod into 3 equal pieces.
Work out how long each piece must be.

C3 Measure these rods.
Work out long 1 piece must be when they are cut.

(a) (3 pieces)

(b) (7 pieces)

(c) (3 pieces)

(d) (6 pieces)

Square Pattern

You need a sheet of plain paper and a calculator.

1 Draw a square on your paper. Make each side 16 cm long.

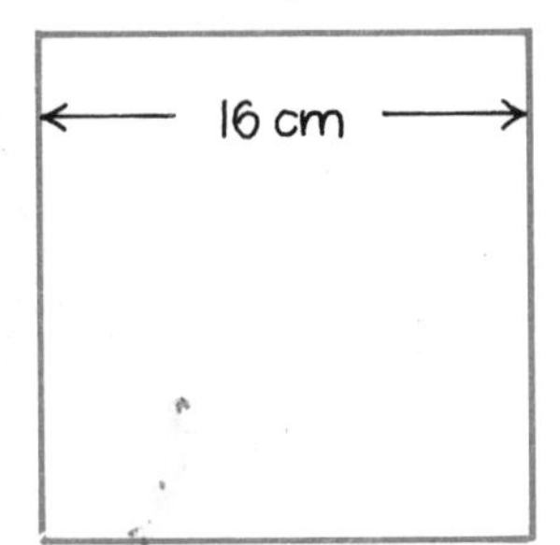

2 Divide each side in 3 equal parts. (Work out $16 \div 3$.) Mark each side.

3 Join the first marks on each side like this.

4 Now join the other marks.

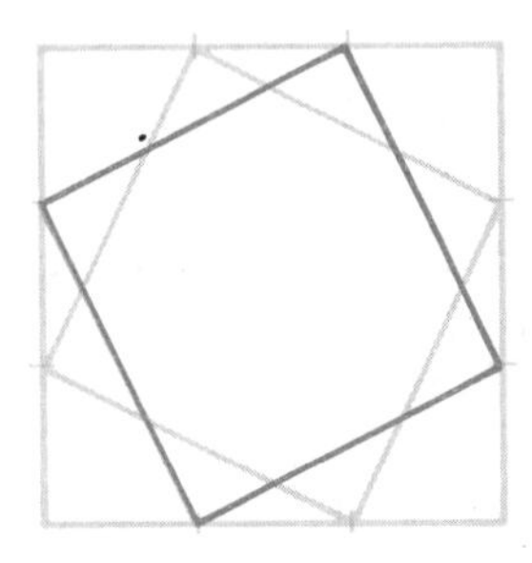

5 Now join the places where the two squares cross.

6 Measure the sides of the new square. Divide the new sides into 3.

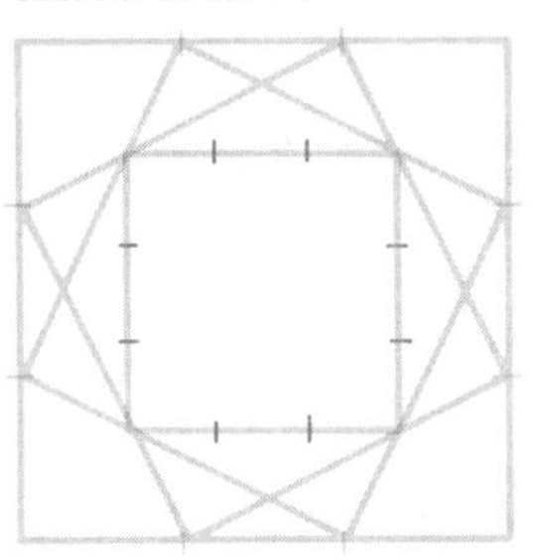

7 Now join the first marks of the new square like in box 3.

Carry on with boxes 4, 5 and 6.

8 Keep on doing this until the middle square is too small to divide up.

Colour the pattern.

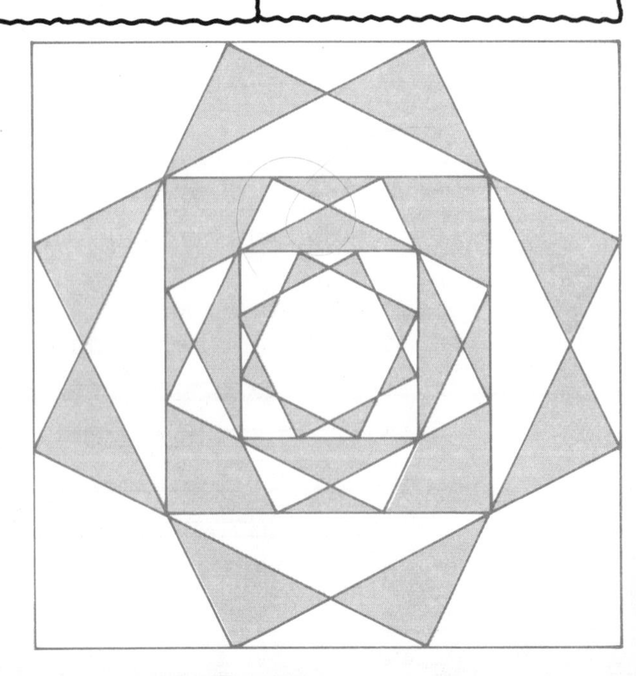

Review: button pressing

In each of these questions
- think carefully whether to use +, −, × or ÷,
- check your answers to see if they are sensible,
- write down the calculation you do.

1 20 people from a youth club go skating.
It costs each of them 80p to get in.
How much do they pay altogether?

2 6 of them go by bike. The rest pay 35p bus fare.
(a) How many go by bus?
(b) How much are the bus fares altogether?

3 It costs 40p to hire skates for one hour.
The groups arrive at 6:00 p.m. and leave at 10:00 p.m.
(a) How many hours does each person need skates?
(b) How much does each person pay for skates?

4 The group buy 5 packets of HOLOs between them.
Each packet has 16 HOLOs in it.
How many HOLOs do they get each?

5 Five of the group go home by taxi.
The taxi fare is £6·85.
They share the fare equally.
How much do they pay each?

6 The five call in at a chippy.

How much do they have to pay altogether?

COD 50p 80p 90p
HADDOCK 65p
PLAICE 70p (when available)
Sausage 38p
Onion rings 24p a portion
CHIPS 20p 35p 45p

Views

A The opposite point of view

Jane serves Jim with three mugs of tea.
She puts them on the counter like this.

A1 Which of these pictures shows the mugs as Jim sees them?

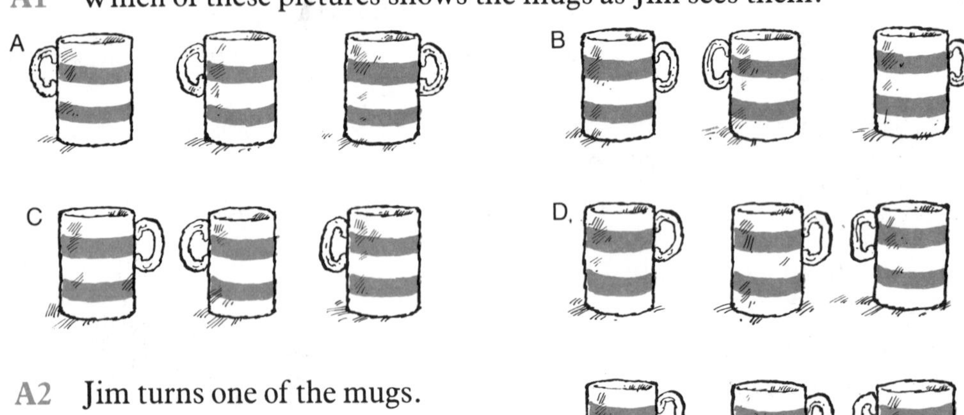

A2 Jim turns one of the mugs.
Now he sees them like this.

(a) Which mug did he turn?

(b) Which of these pictures shows the mugs as Jane sees them now?

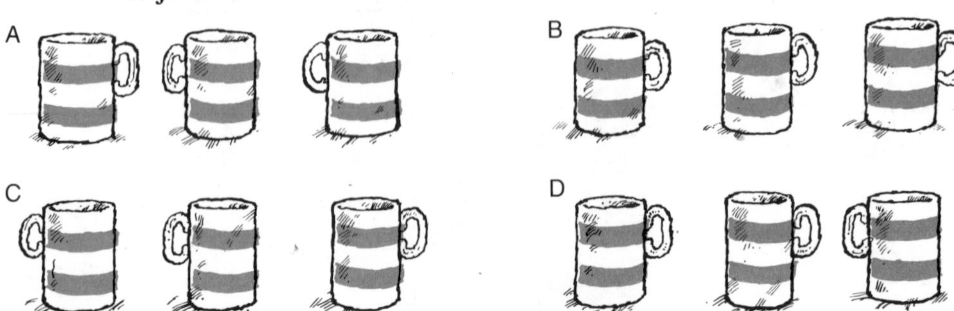

Here is a place
correctly set for school dinner.

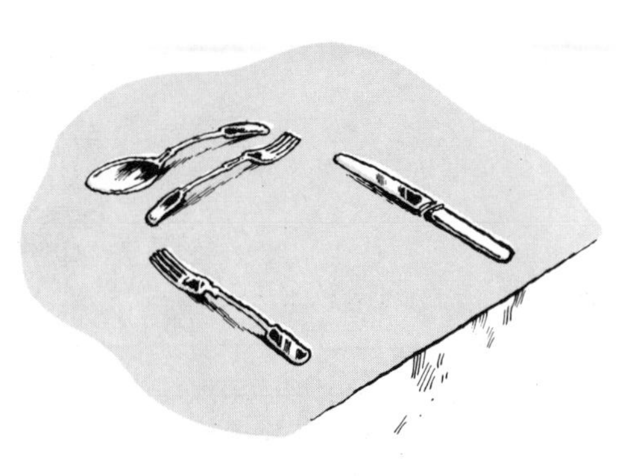

Fred sets some places for school dinner.
He leans across the table
to set the places on the far side.
Here is the table after Fred set it.

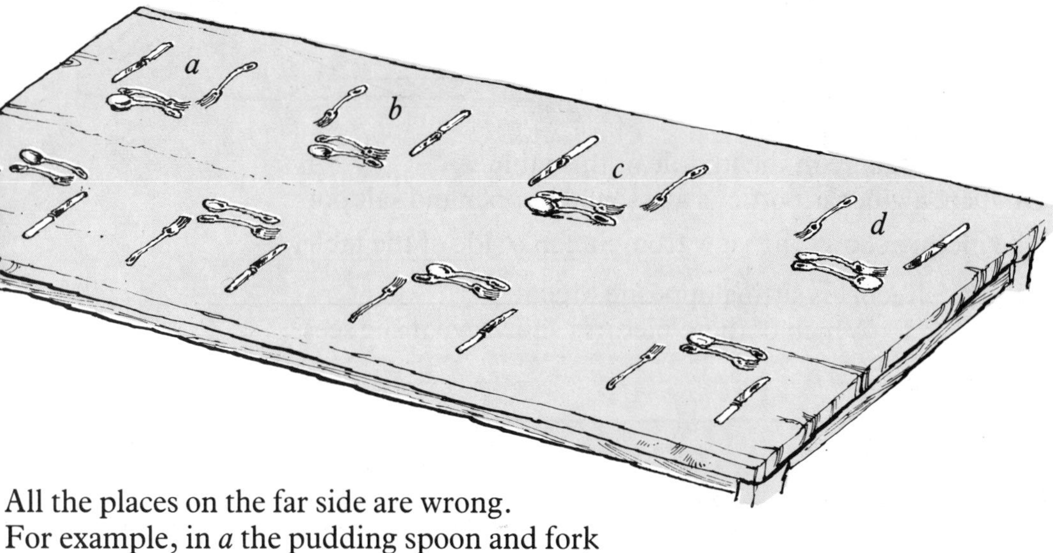

All the places on the far side are wrong.
For example, in *a* the pudding spoon and fork
both point the wrong way.

A3 Write down what is wrong in settings *b*, *c* and *d*.

A4 Sally sits opposite Daren at dinner.
She puts a salt pot, a pepper pot
and a knife and fork
in front of her like this.

Which of these pictures show what Daren sees?

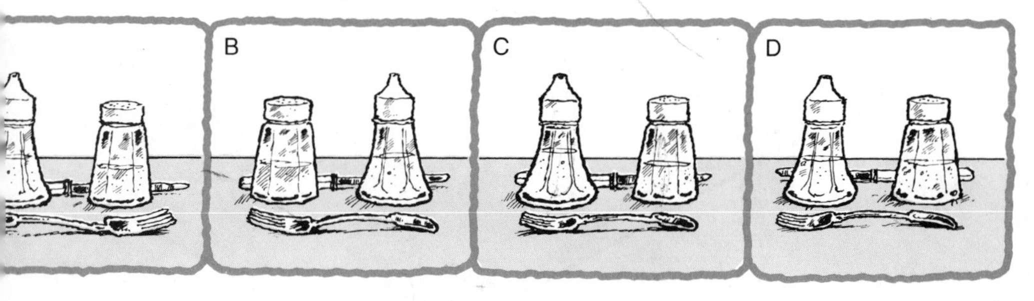

B All square

There is a square in the middle of this table.
On it are a vinegar bottle, a glass, and pepper and salt pots.
The picture shows the view from Stefan's side of the table.

B1 Rachel is sitting opposite Stefan.
(a) Which of these pictures shows what she sees?

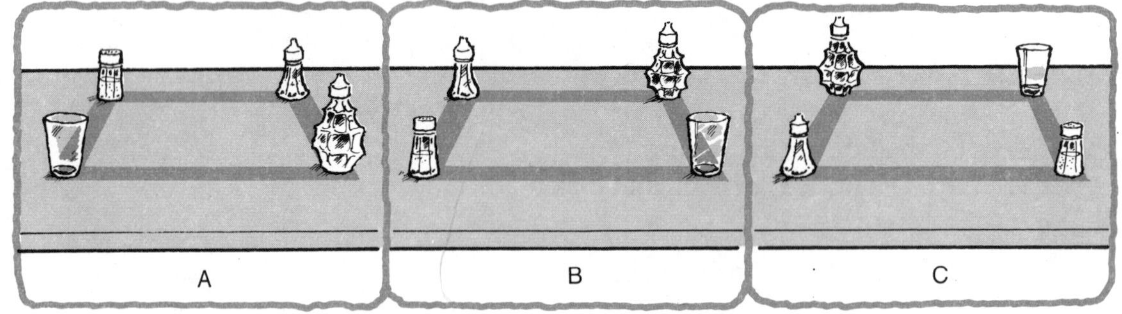

(b) Which picture shows what Melinda sees?
(c) Which picture shows what Joseph sees?

B2 In the picture at the top of the page the corners of the table are numbered.
Here is the same table.
What number is in the corner marked **?**?

Here are four pictures.
Three pictures show what you see
if you stand at one of the corners of the table.
One picture is wrong. The glass, bottle and pots
are shown on the wrong corners.

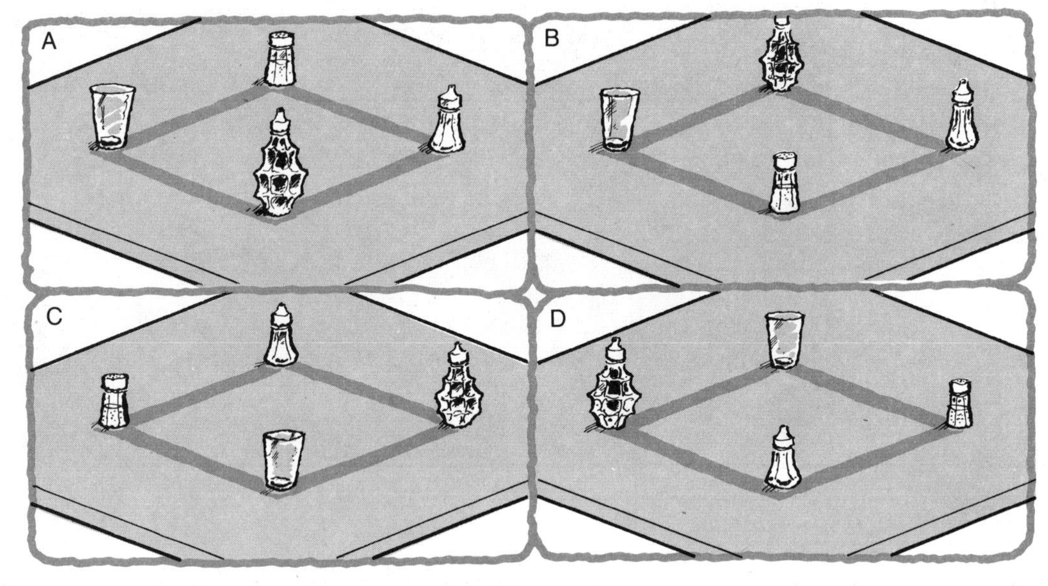

B3 Which is the wrong picture?

B4 Which corner are you looking from in the pictures
that are right?
Write your answers like this.

Picture . . . is the view from corner . . .

B5 Melinda swaps over the two objects nearest to her.
Which two objects are nearest to Rachel now?

B6 Which of these pictures show what Joseph sees now?

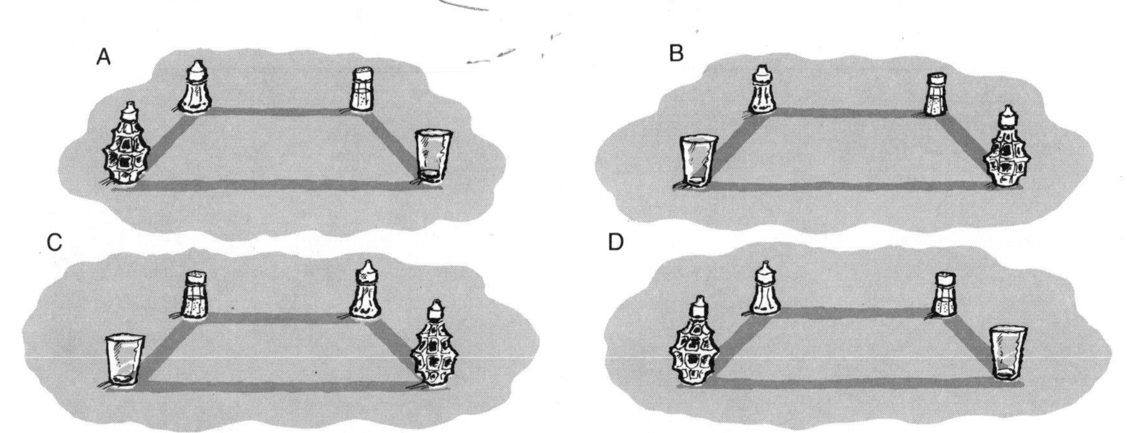

C Views – front, back, side and top

This is a photograph of a 1938 Mercedes Benz W154.

You can see a lot of detail in the photograph,
but some things cannot be seen.
You cannot see what shape the back of the car is;
you cannot even see how many wheels it has!

Imagine the car is put into a glass box.

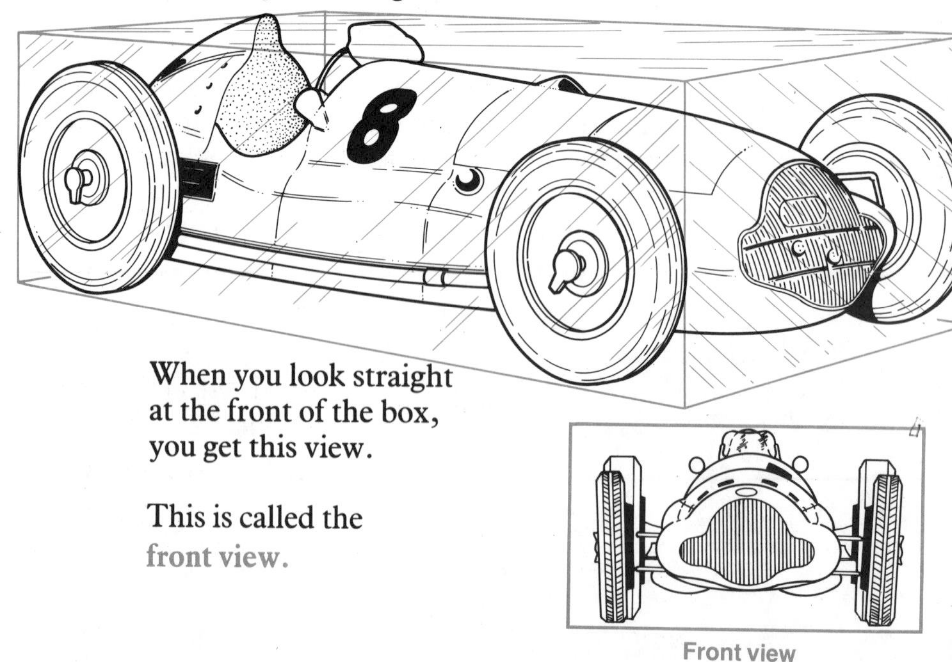

When you look straight
at the front of the box,
you get this view.

This is called the
front view.

Front view

If you look straight at the side of the box, you see a **side view**.

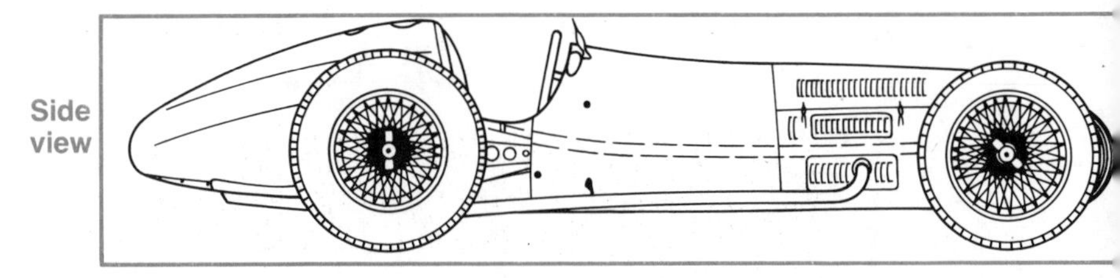

In the same way this is the **back view**.

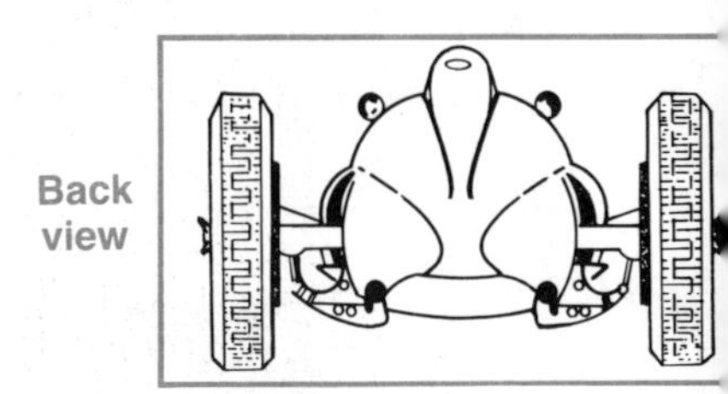

This is the **top view**. The top view is often called the **plan view**.

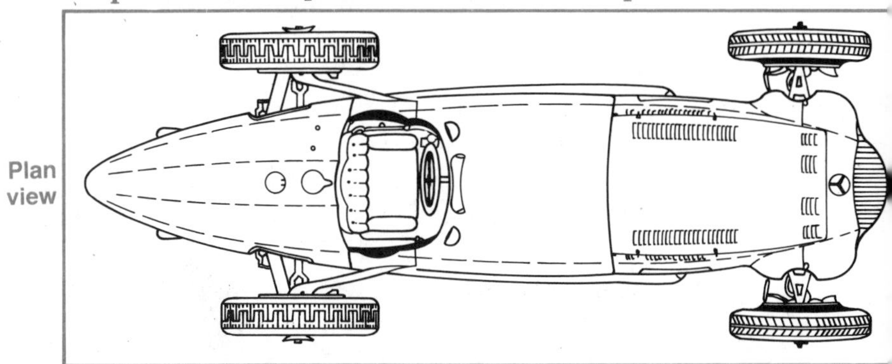

C1 Which view shows most clearly how many wheels the car has?

C2 (a) Do the wheels have spokes?
(b) Which view helps you answer this question?

C3 Which view shows most clearly that the front tyres are smaller than the back tyres?

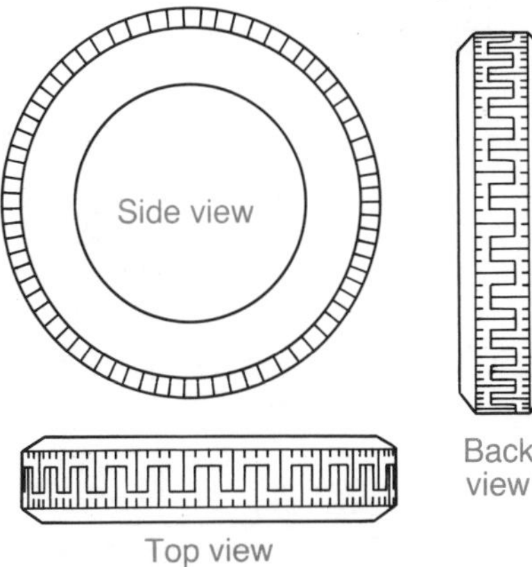

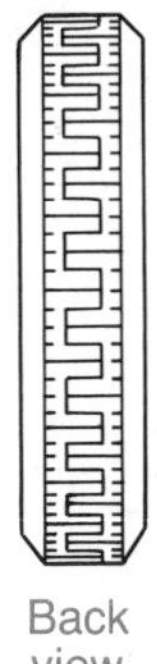

C4 Here are three views of a spare back t
(a) What do you notice about the top view and the back view?
(b) Which of these three views woul the front view look like?

C5 These are photos of three old windmills.
There are views of the windmill below.
Which view goes with which windmill?
Write your answers like this *A is the plan view of Packenham.*

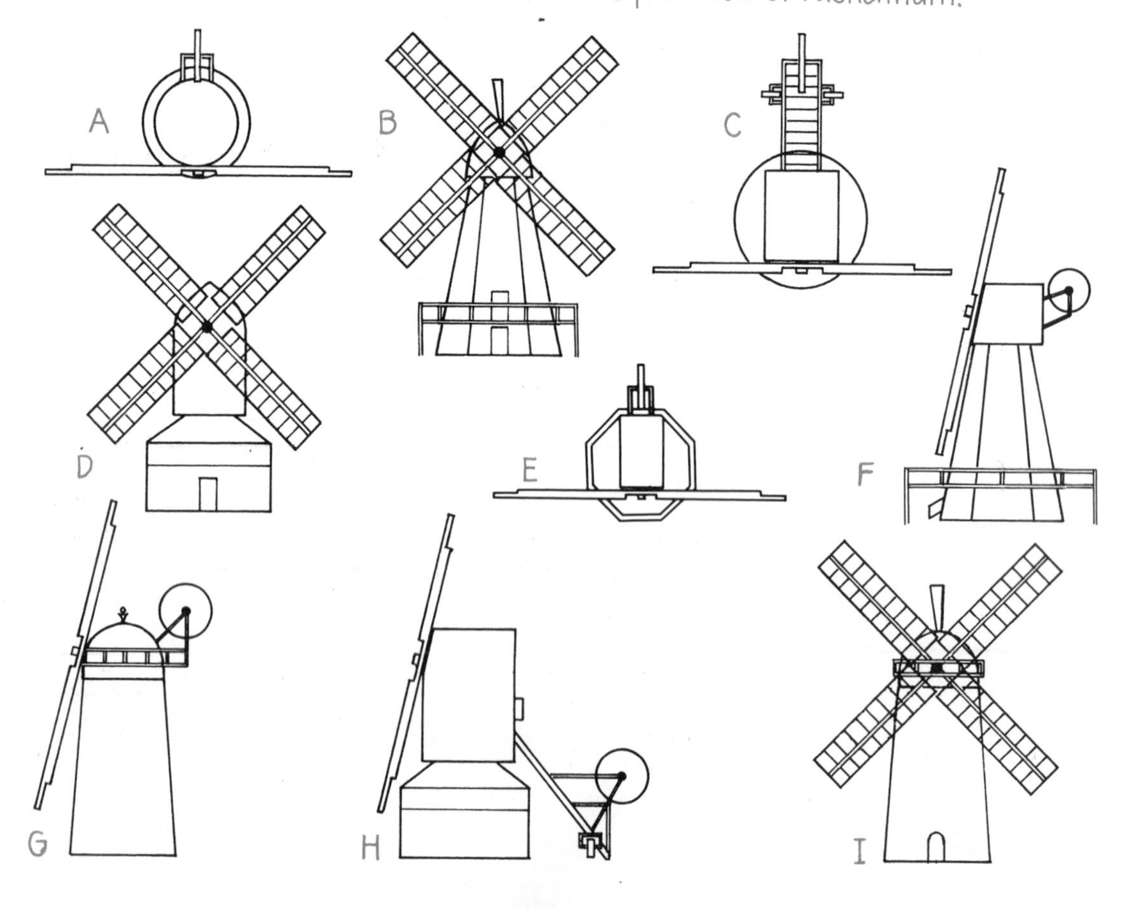

C6 Here are some views of well-known objects.
Can you name them?
There are two views of each object here.

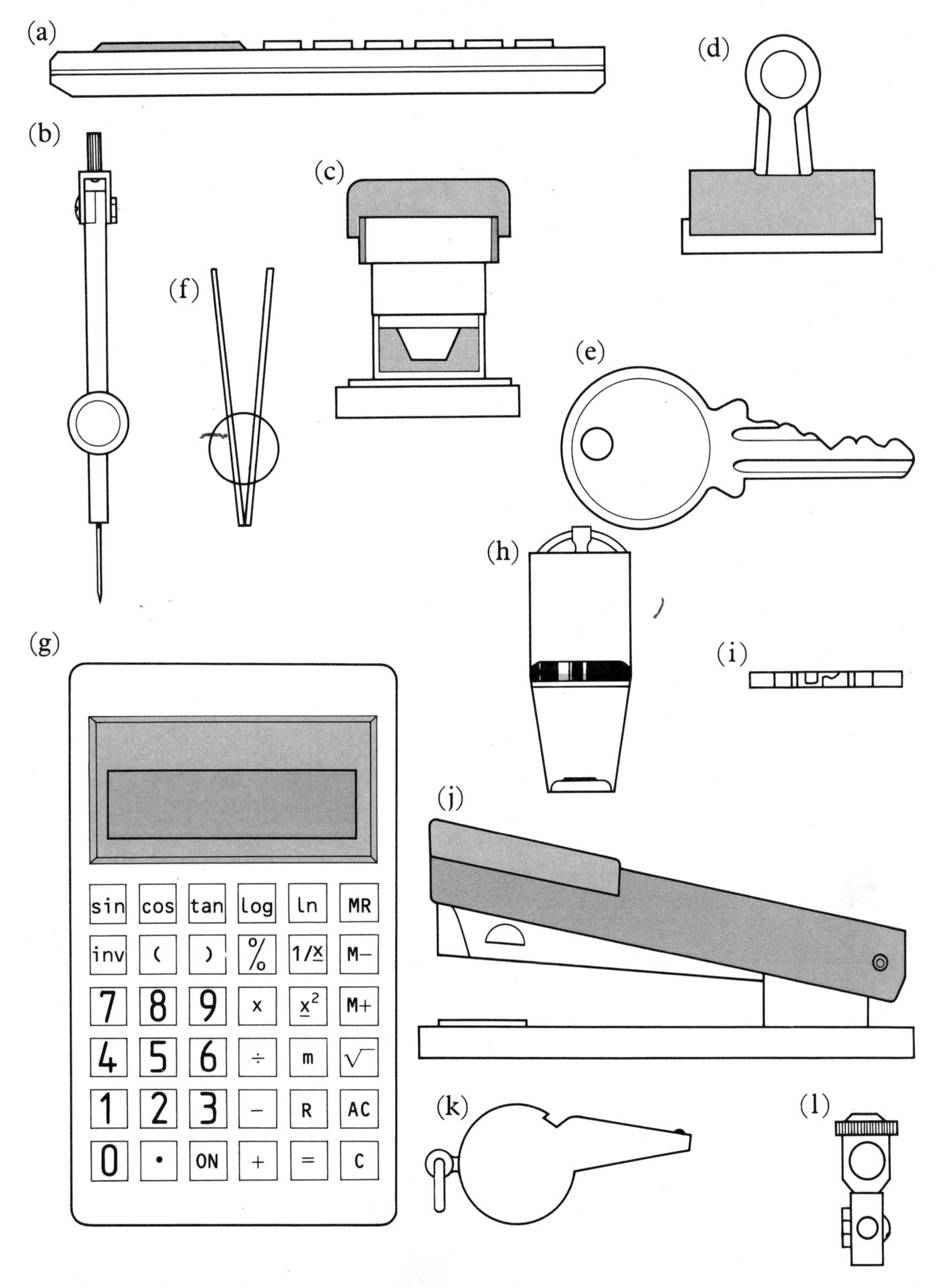

D Look at it this way

You need worksheet G4–2, scissors and glue.

Cut out the model of the shed.
Fold it, and stick it together.

The front of the shed
has the door in it.

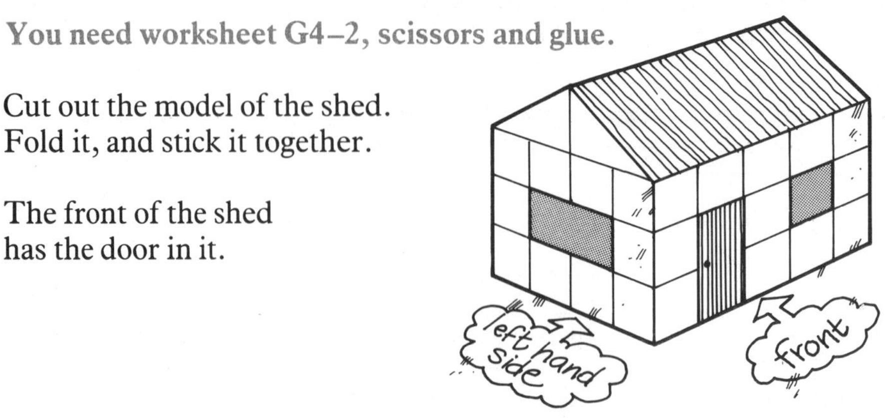

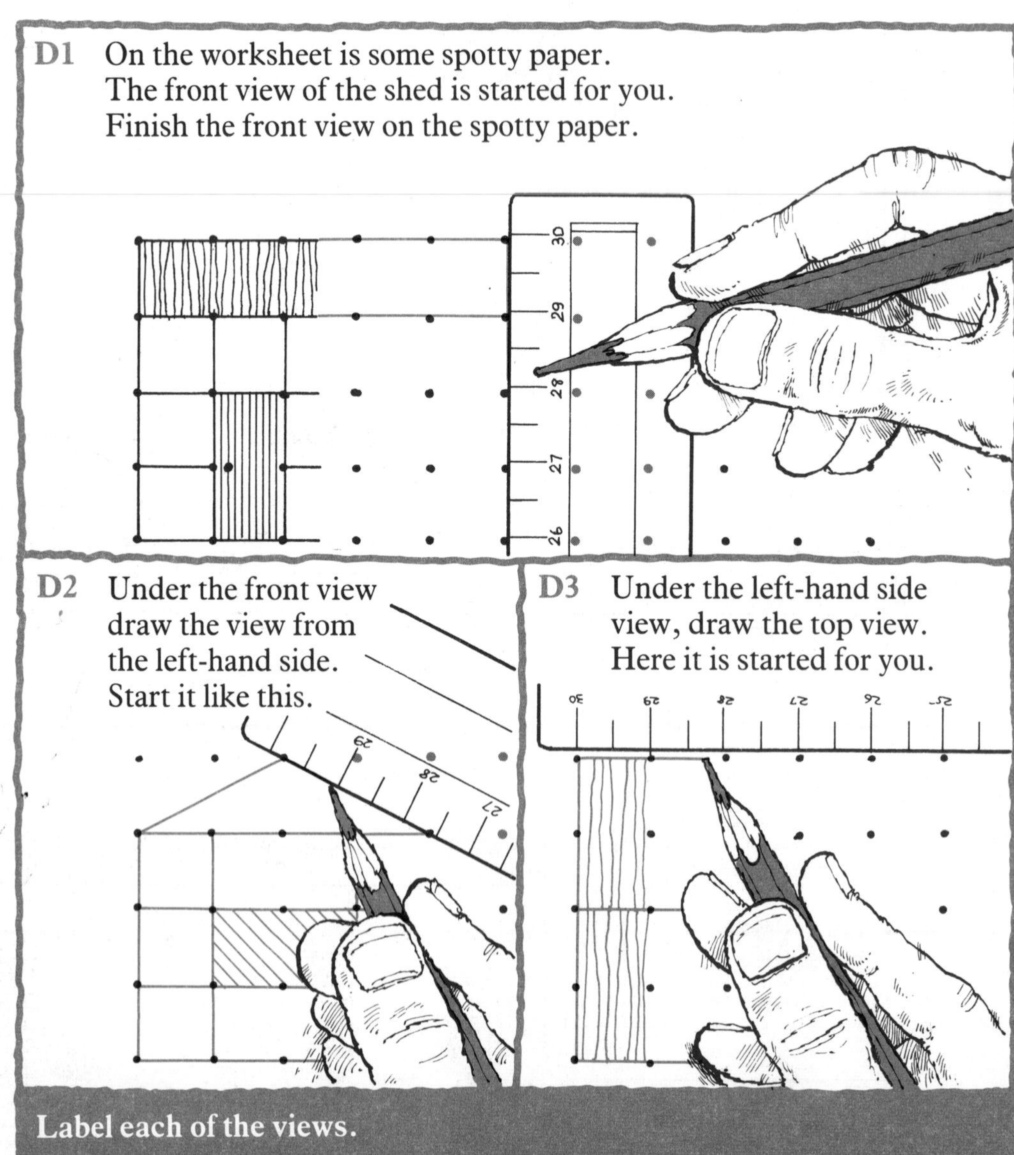

D1 On the worksheet is some spotty paper.
The front view of the shed is started for you.
Finish the front view on the spotty paper.

D2 Under the front view draw the view from the left-hand side.
Start it like this.

D3 Under the left-hand side view, draw the top view.
Here it is started for you.

Label each of the views.

D4 On the spotty paper, draw the back view of the shed.
Draw it next to the front view.

Label the back view.

D5 Next to the left-hand view, draw the right-hand view.
Label it carefully.

Keep the model shed safe. You will need it later.

You need worksheet G4–3.

1 Cut out the model house.
Fold along the dotted lines.

2 If you want to colour the house, do it now.

3 Now stick the chimney together like this.

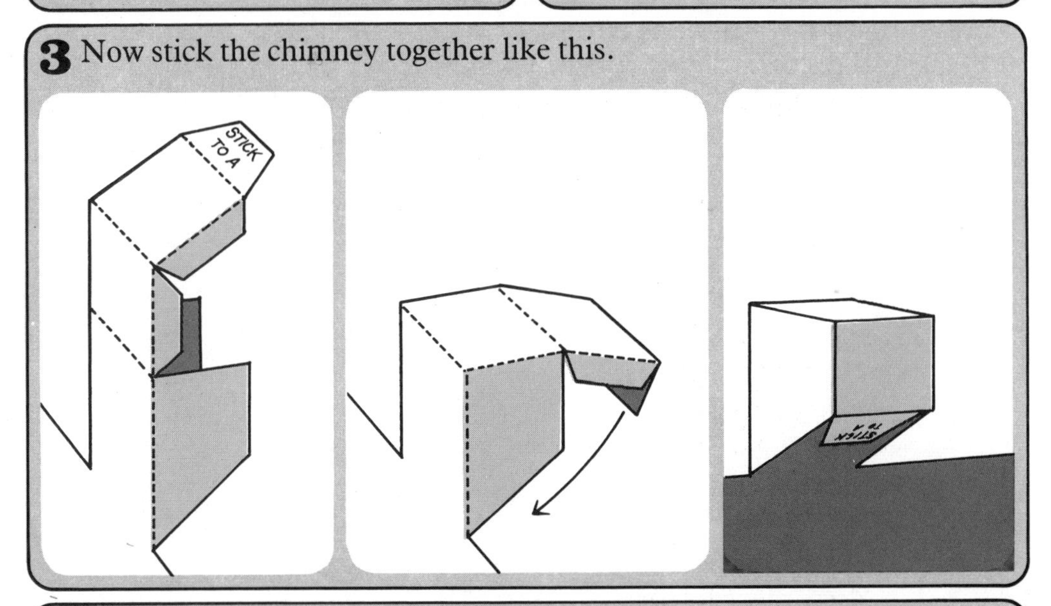

4 Stick the rest of the model together.
Stick the chimney to the roof at A.

D6 Which of these is a photo of the house?

A

B

You need worksheet G4–4.

One of the doors of your house has a letterbox.
Stand your model on the rectangle marked 'House'.
Make sure the door with the letterbox is in the right place.

D7 There are 8 arrows on the worksheet, marked A to H.
Here are 8 views of the house.

Which arrow goes with which view?

Write your answers like this.

Arrow A goes with view ...

Try to answer without looking round the model.
Then check by looking round.

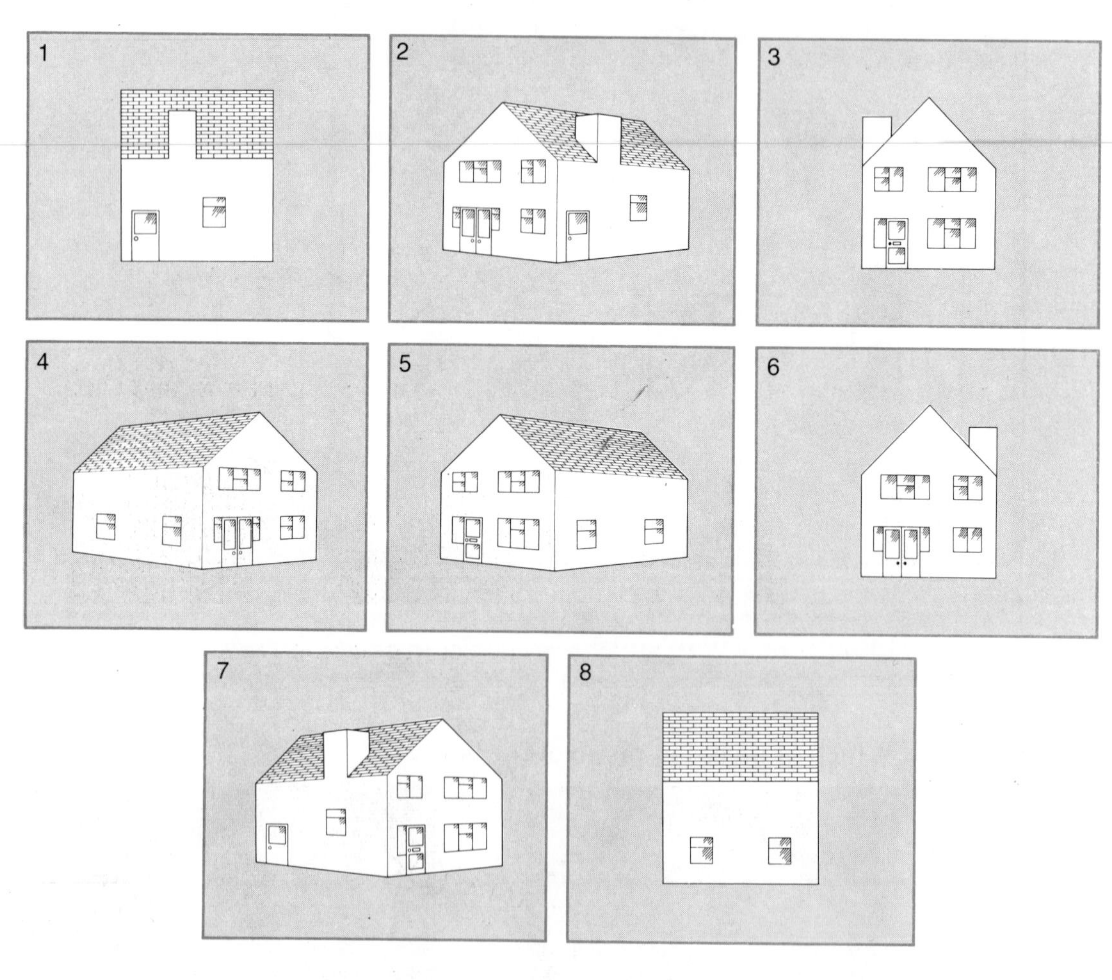

D8 Draw a plan view of your house.
You will have to do some measuring.
If you want, do it on spotty paper.

Keep your house on worksheet G4–4.

On the worksheet, there are 2 rectangles where you can put the shed.
Put your model shed on the shaded rectangle.
Make sure the door is in the right place.

D9 Here are some views of the house and shed.
Which arrow (A, B, C and so on) goes with which view?

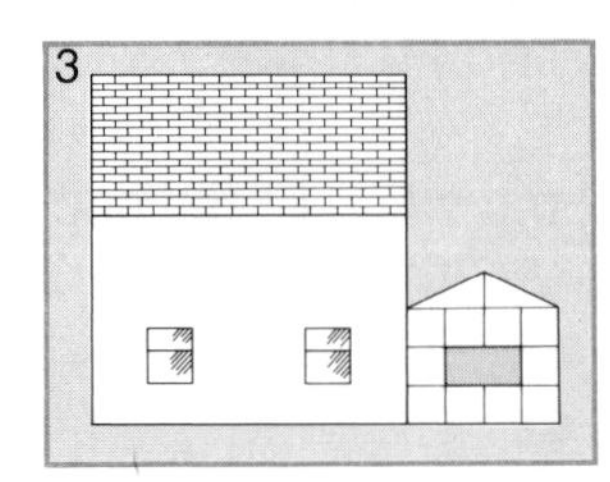

D10 From the direction of one arrow, you cannot see the shed.
Which arrow is this?

D11 In these views, you can only see the outline of the house and shed.
One view is wrong. Which view is that?

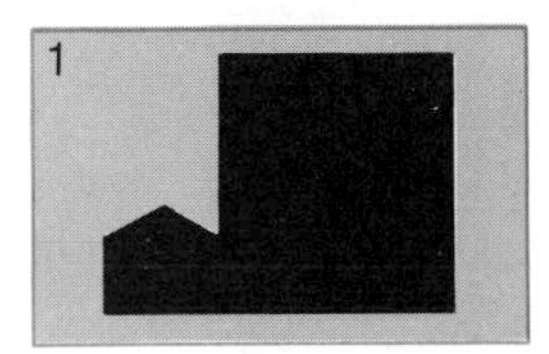

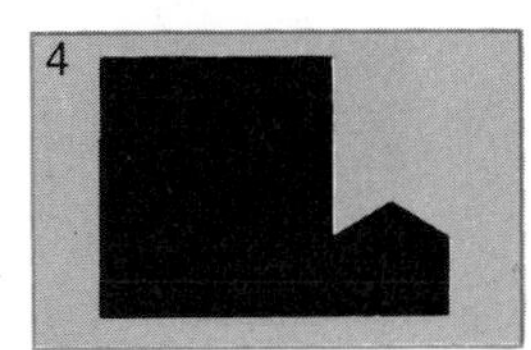

D12 Look at the views which are correct in question D11.
Which arrow goes with which view?

D13 Which of these **plan views** are correct?

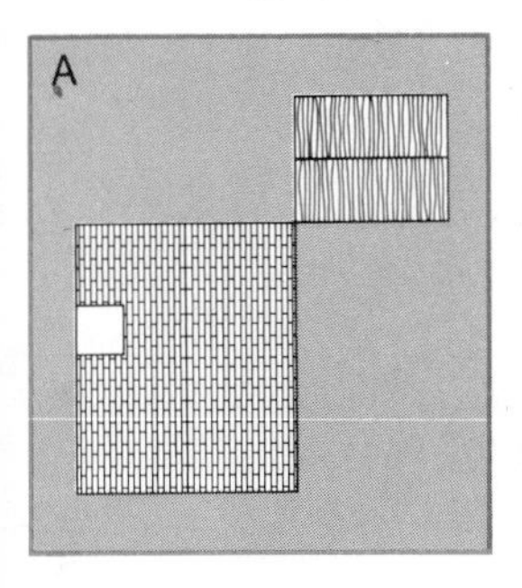

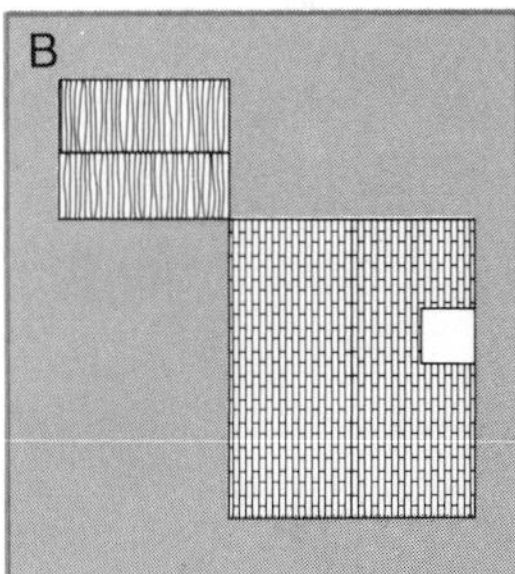

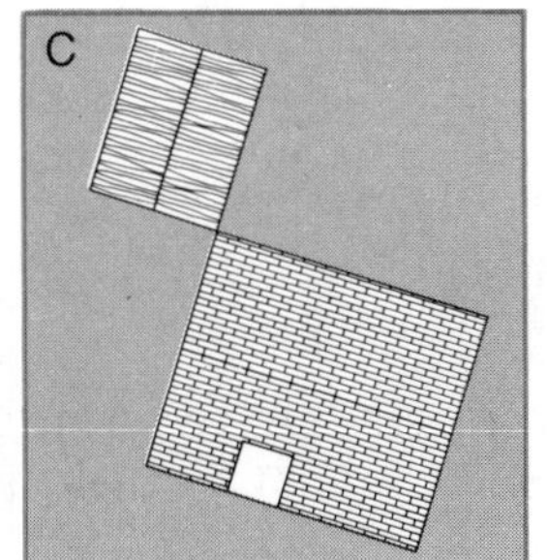

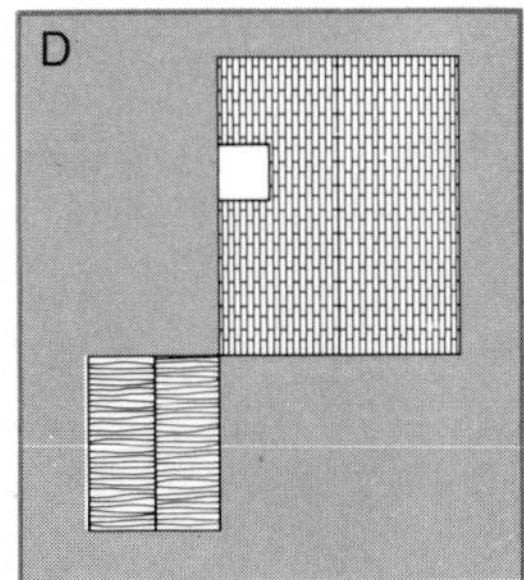

Move the shed to the other rectangle.

D14 From which direction can you **not** see the shed now?

D15 This is the view from A when the shed is on the **shaded** rectangle.

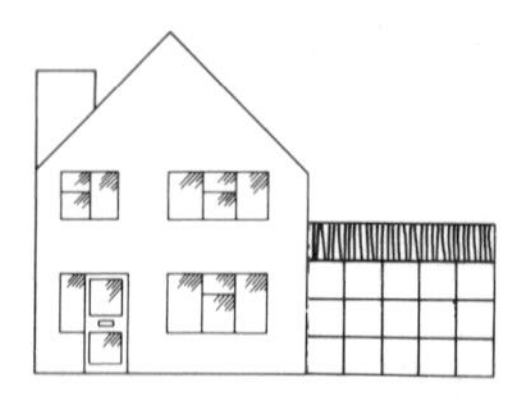

This view is still the same now the shed is on the new rectangle. Check that you can see this.

The view is still the same in one other direction. Which direction is this?

D16 Here are some views of the house and the shed. In some views, the shed is on the shaded rectangle. In some views, it is on the unshaded one.

For each view, write *shaded* or *unshaded*

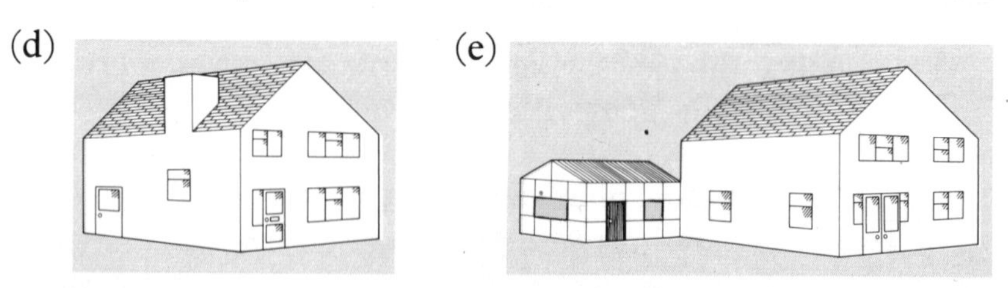

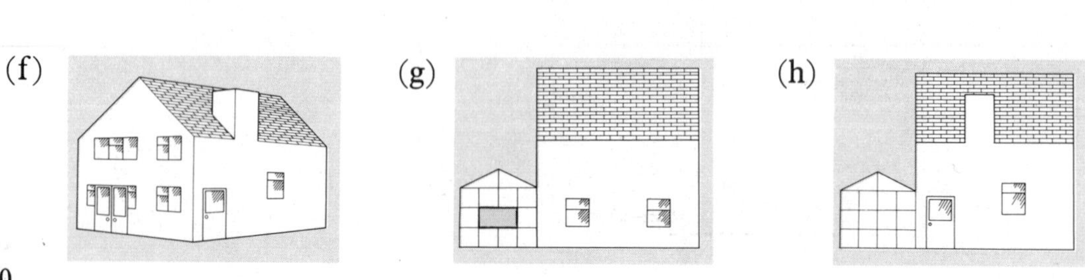

Review: decimals

1 Find the lengths marked **?**.

(a)

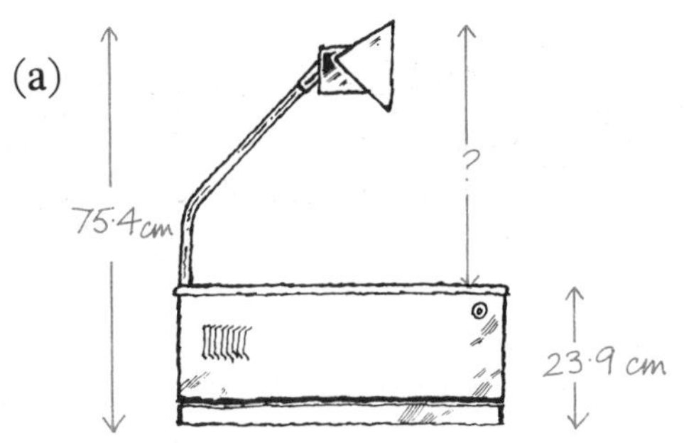

(b)

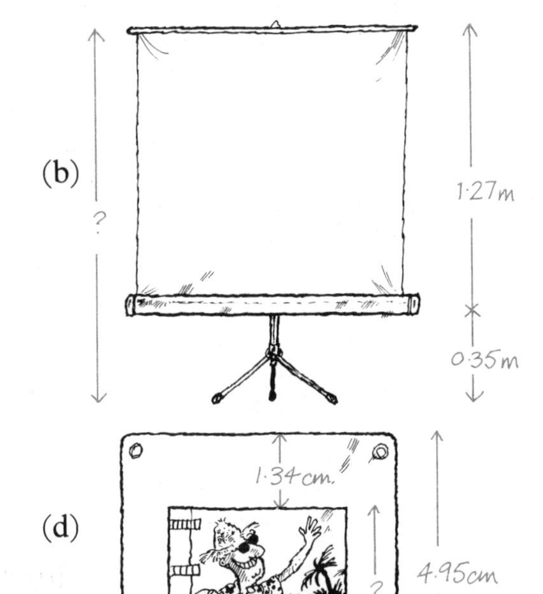

(c)

?
0·06m
0·37m

(d)

1·34cm
?
4·95cm

2 Try to work these out in your head. If you can't, use pencil and paper. Then check with a calculator.

(a) $3 - 2{\cdot}15$ (b) $5 + 3{\cdot}71$ (c) $3{\cdot}42 - 2$

(d) $5{\cdot}82 + 6$ (e) $12 - 9{\cdot}37$ (f) $16 + 4{\cdot}83$

3 Find the lengths marked **?**.

(a)

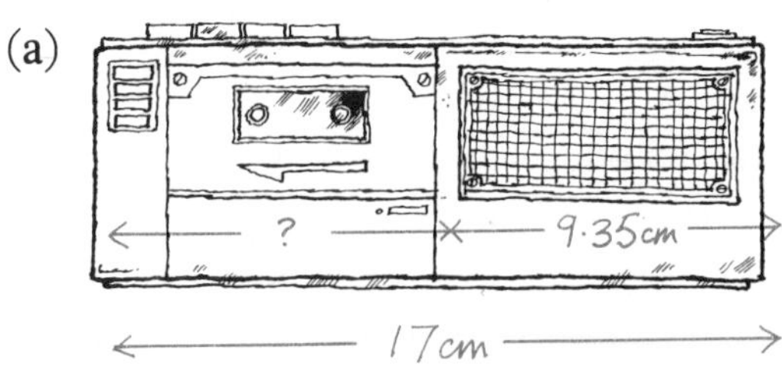

(b)

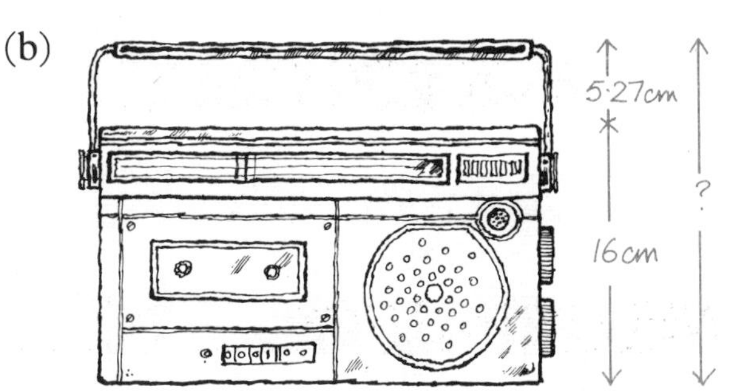

(c)

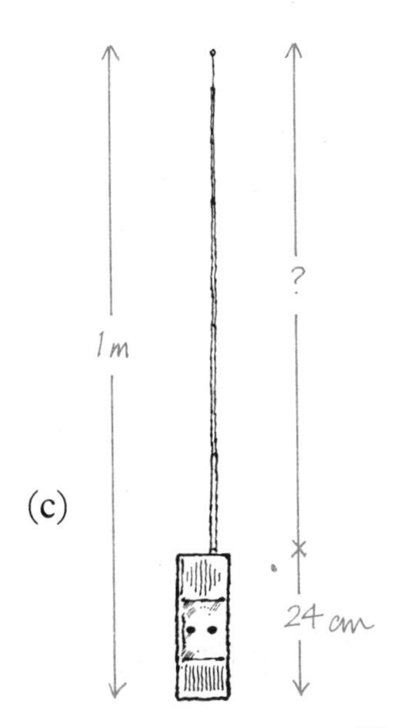

7 Working out percentages

A 10 per cent

New Lick Sticks are 10% longer than the old ones.
10% is the same as 1 tenth.

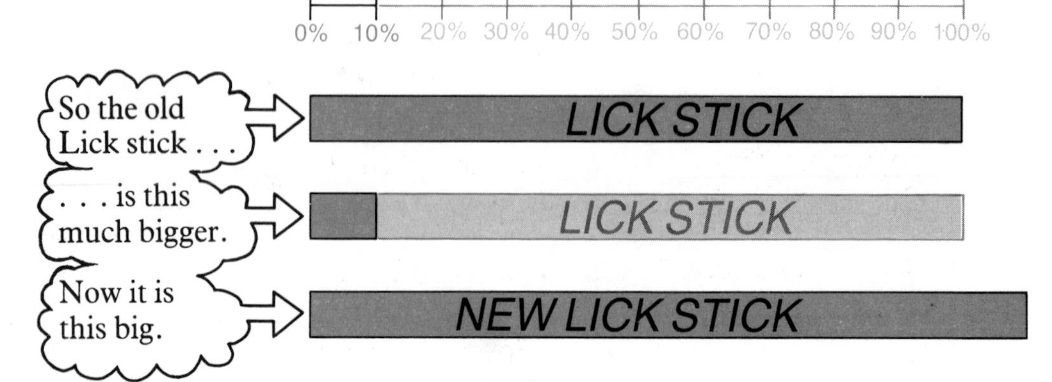

The old Lick Stick was 90 mm long.
1 tenth of 90 is 9,
so 10% of 90 mm is 9 mm.

The new stick is 99 mm long.

You can think of it like this.

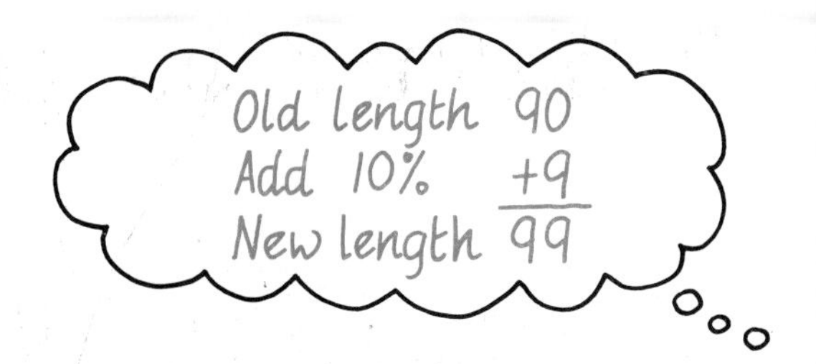

A1 The old Lick Stick used to weigh 50 grams.
The new one is 10% heavier.

How much does the new one weigh?

A2 The old Lick Stick used to cost 30p.
The new one is 10% dearer.

How much does the new one cost?

A3 The old Lick Sticks come in boxes of 200.
The new boxes have 10% more Lick Sticks in them.
How many Lick Sticks are in the new boxes?

A4 The Cambridge English Dictionary claims it is '10% bigger in every way'.

(a) The old one was 160 mm wide.
How wide is the new one?

(b) The old one was 230 mm tall.
How tall is the new one?

(c) The old one had 350 pages.
How many has the new one?

(d) The old one cost £10.
How much is the new one?

A5 The old Cambridge Dictionary had 800 printing errors!
How many will the new one have, if it has 10% more?

A6 This is a leaflet for the *Viking*.

The *New Viking* is 10% longer, wider, faster and dearer!

Write out a leaflet for the *New Viking*.

Anglo-Irish Ferries	VIKING leaflet
Length	190 m
Width	20 m
Speed	30 knots
One way crossing	£20
Return fare	£40
Special family return	£120

B Money

B1 To find 10% of £2, you can think of £2 as 200 pence.
Work out 10% of £2.

There is a different way to work out money percentages.

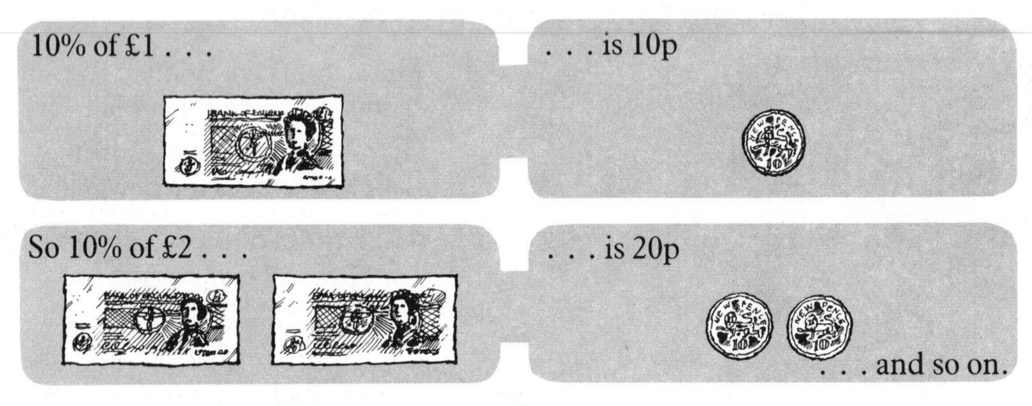

10% is the same as 10p in the pound.

B2 What is 10% of £7?

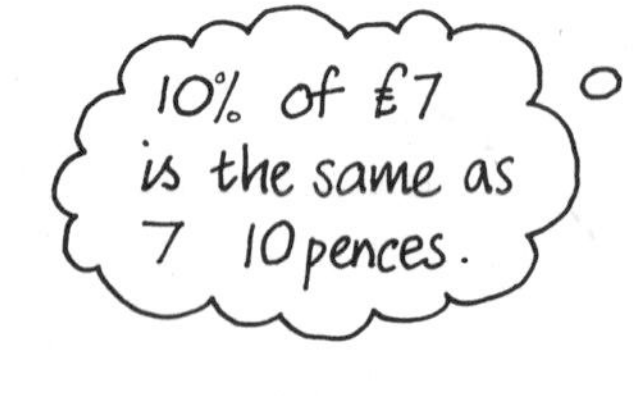

B3 Write down

(a) 10% of £6 (b) 10% of £8

(c) 10% of £12 (d) 10% of £23

B4 The boots in the window cost £21.
The deposit on the boots is 10% of £21.

(a) How much is the deposit on the boots?

(b) How much are the deposits on the other articles?

This shopkeeper is putting up all her prices by 10%.

The peat used to cost £6.

10% of £6 is the same as 6 tenpences.

So the peat's price goes up 60p.
The new price is £6·60.

B5 The watering can used to cost £5.

(a) What is 10% of £5?

(b) What does the can cost when it is 10% dearer?

B6 The shade used to cost £18.
How much is it when it is 10% dearer?

B7 What is the new price of

(a) the ladder (b) the barrow (c) the table

This shopkeeper is having a sale.
He is taking 10% **off** all his prices.

The doll used to be £4.

10% of £4 is the same as 4 tenpences.

So he takes off 40p.

The doll now costs £3·60.

B8 The jigsaw used to be £5.

(a) What is 10% of £5?

(b) How much is the jigsaw when you take off 10%?

B9 After you take off 10%, how much is each of these?

(a) The *Action Book* (b) The teddy bear

(c) The racing car (d) The train set

C Other percentages

If you can work out 10% then you can easily work out 20%.

The skirt costs £4, so the deposit is now 20% of £4.

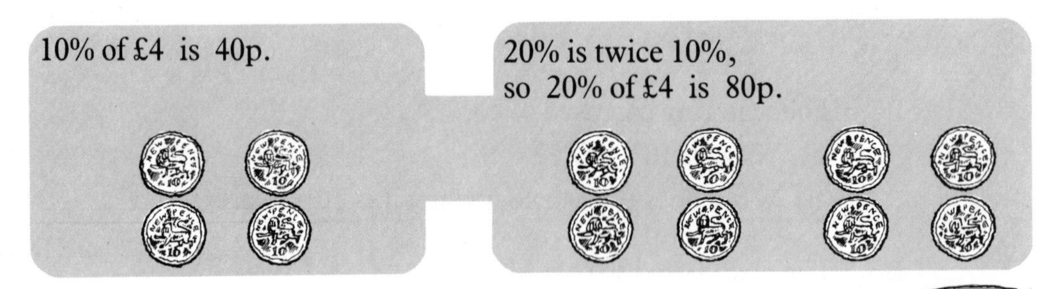

10% of £4 is 40p.

20% is twice 10%, so 20% of £4 is 80p.

C1 The deposit on this bag is 20% of £8.

(a) What is 10% of £8?

(b) What is 20% of £8?

C2 What will the deposit be on each of these?

C3 Work out 20% of

(a) £3 (b) £21 (c) £30

(d) £100 (e) £17 (f) £27

C4 20% is the same as 20p in the pound. So 20% of £6 is the same as 6 lots of 20p.

Use this method to work out

(a) 20% of £7 (b) 20% of £5

(c) 20% of £9 (d) 20% of £14

(e) 20% of £2 (f) 20% of £16

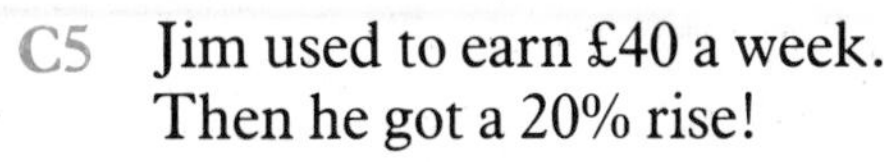

C5 Jim used to earn £40 a week.
Then he got a 20% rise!

(a) What is 20% of £40?

(b) What is Jim's new wage?

C6 This coat was £25.
Ajit got 20% off in the sale.

(a) What is 20% of £25?

(b) How much did Ajit pay?

C7 The Jupiter Bar was 60 mm long.
Now it is 20% longer.

(a) What is 10% of 60 mm?

(b) What is 20% of 60 mm?

(c) How long is the Jupiter Bar now?

C8 P-nuts used to come in 80 gram packs.
Now the packs are 20% heavier.
How heavy are they now?

C9 30% is 3 times 10%.
With money, you can think of 30% as 30p in the pound.

Use any method you want to work out

(a) 30% of £6 (b) 30% of £9 (c) 30% of £3

(d) 30% of £12 (e) 30% of £10 (f) 30% of £15

C10 How much do each of these cost after taking 30% off?

(a)

(b)

(c)

C11 5% is half of 10%. What is

(a) 5% of £6 (b) 5% of £16 (c) 5% of £12

C12 Bobby's Store takes 5% off if you pay cash.
How much do each of these cost if you pay cash?

(a)

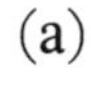

(b)

(c)

Review: chapters 2 and 3

1 George makes his Christmas pudding using this recipe.

> **Inexpensive Christmas Pudding**
> serves 12
>
> | | |
> |---|---|
> | 125 g plain flour | 150 g peel |
> | 125 g breadcrumbs | 125 g castor sugar |
> | 25 g SR flour | 1 lemon (rind and juice) |
> | 250 g suet | 2 eggs |
> | 250 g sultanas | Milk to mix |
> | 250 g raisins | |
>
> Mix dry ingredients. Stir in beaten eggs, and milk to make moist. Steam in well greased pudding basin for 4 hours.

(a) He needs 125 g of plain flour. What is 125 g in kg?

(b) He opens a 2 kg bag of plain flour. How much is left after taking 125 g out?

(c) He uses 'mixed dried fruit' instead of the sultanas and raisins. How many grams does he use?

(d) He opens a 1·1 kg packet of mixed dried fruit. How many kg are left after he makes the pudding?

(e) How many kg do the 'dry ingredients' weigh altogether?

(f) A small egg weighs about 45 g. A large egg weighs about 70 g. About how much do you think George's 2 eggs will weigh?

(g) About how much will his pudding weigh altogether?

(h) The pudding serves 12. About how many grams is that each?

2 Ruth makes a pudding for 48! Write out the list of ingredients and the weights Ruth uses. Use kg and grams.

3 Jim uses the recipe to make a pudding for 4 people.

(a) What must he divide the quantities by?

(b) How much peel will he need?

(c) **Roughly** how much suet will he need?

(d) About how heavy will his pudding be?

(e) Jim puts 8 old sixpences in his pudding. How many should George put in?

(f) Jim's guests take 15 minutes to eat his pudding. How long will George's take?

Review: chapters 4 and 5

1 6 friends go to London for the day.
This is the timetable from Blagdon.

(a) What time does the first train for London leave Blagdon?

(b) Could you get a hot breakfast on that train?

(c) The friends want to get to London by 12 o'clock.
Which is the best train to catch?

(d) The train is 25 minutes late.
What time does it arrive in London?

(e) How long did their train take altogether from Blagdon to London?

(f) They were meeting a friend at the station at a quarter to 12.
How long did the friend have to wait

To LONDON

Blagdon dep.		London arr.
0712	☕	0905
0936	☕ ‖	1140
1040		1237
1156	☕	1341
1359		1550
1700	☕	1903
1710	☕ ‖	1900
1745	☕	2001
1805		2010
1907		2115
2001		2316

☕ Buffet car
‖ Restaurant car

2 The 6 paid a special 'group price'.
It cost them £20 return.
How much will they each have to pay?

3 They go to a laser show.
They have to pay for 3 adults and 4 under 14's.
They split the cost equally between them.
How much do they each pay?
(Remember there are 7 friends altogether.)

4 They buy a raffle ticket between them and win £10!
How much can they have each?

5 The last three trains back to Blagdon leave London at 20:01, 21:05 and 23:15.
They get to the station at 11 p.m.
What time train can they catch?

6 Their train is timetabled to get to Blagdon at 03:10. It is $1\frac{1}{2}$ hours late.
What time do they get in?

Review: chapters 6 and 7

Here are pictures of two pleasure boats.

Starlight

Crystal light

1 These are the plan, side and front views of the boats.
Which are views of *Starlight* and which are of *Crystal light*?

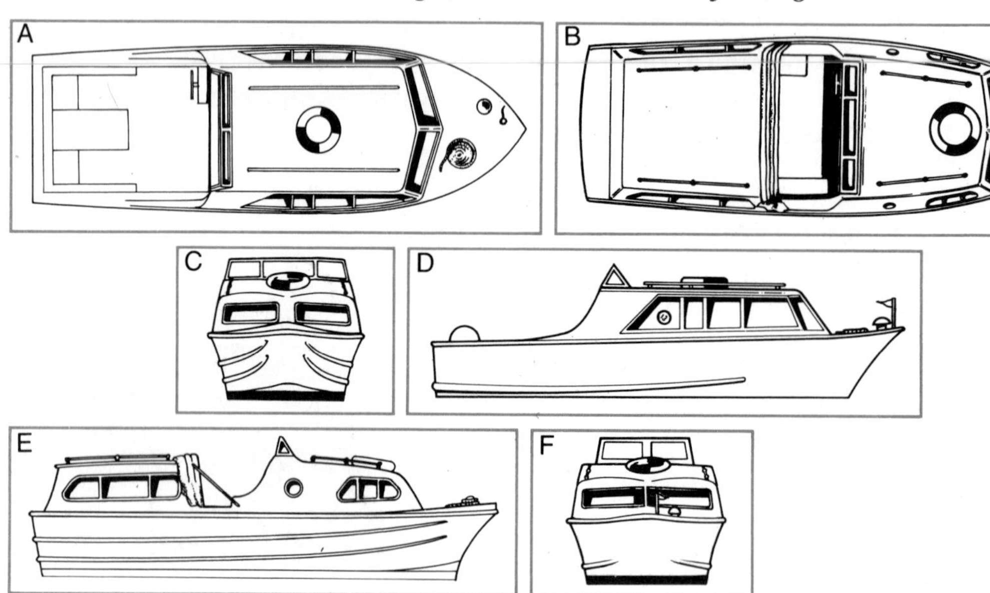

2 *Starlight*'s headroom is 200 cm.
Crystal light's headroom is 10% less than *Starlight*'s.
What is *Crystal light*'s headroom?

3 This is part of *Starlight*'s fact-sheet.
Crystal light is 20% longer
and 20% wider than *Starlight*.
Write out a fact-sheet
for *Crystal light*.

Starlight	
Length	700 cm
Beam (width)	250 cm
Headroom	200 cm